THE BI
OUT OF TH

Homeopathy was founded in 1796 by the German physician Samuel Hahnemann, who ardently proposed that "like cures like," counter to the conventional treatment of prescribing drugs that have the opposite effect to symptoms.

Alice A. Kuzniar critically examines the alternative medical practice of homeopathy within the Romantic culture in which it arose. In *The Birth of Homeopathy out of the Spirit of Romanticism*, Kuzniar argues that Hahnemann was not an iconoclast and visionary, but rather a product of his time with links to his contemporaries such as Goethe and Alexander von Humboldt. It is the first book in English to examine Hahnemann's unpublished writings, including case journals and self-testings. Kuzniar's engaging writing style seamlessly weaves together medical, philosophical, semiotic, and literary concerns in order to reveal homeopathy as a phenomenon of its time. *The Birth of Homeopathy out of the Spirit of Romanticism* sheds light on issues that continue to dominate the controversy surrounding homeopathy to this very day.

ALICE A. KUZNIAR holds a University Research Chair at the University of Waterloo and is a professor in the Department of Germanic and Slavic Studies.

THE BIRTH OF HOMEOPATHY
—out of the—
SPIRIT OF ROMANTICISM

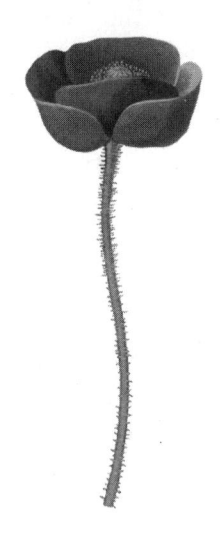

Alice A. Kuzniar

UNIVERSITY OF TORONTO PRESS
Toronto Buffalo London

© University of Toronto Press 2017
Toronto Buffalo London
www.utppublishing.com
Printed in the U.S.A.

ISBN 978-1-4875-0117-4 (cloth)
ISBN 978-1-4875-2126-4 (paper)

∞

Printed on acid-free, 100% post-consumer recycled paper
with vegetable-based inks.

Library and Archives Canada Cataloguing in Publication

Kuzniar, Alice A., author
The birth of homeopathy out of the spirit of Romanticism /
Alice Kuzniar.

Includes bibliographical references and index.
ISBN 978-1-4875-0117-4 (cloth). – ISBN 978-1-4875-2126-4 (paper)

1. Hahnemann, Samuel, 1755–1843. 2. Homeopathy – Germany – History.
3. Alternative medicine – Germany – History. 4. Romanticism – Germany.
I. Title.

RX51.K89 2017   615.5'320943   C2016-907267-3

This book has been published with the help of a grant from the Federation for the Humanities and Social Sciences, through the Awards to Scholarly Publications Program, using funds provided by the Social Sciences and Humanities Research Council of Canada.

University of Toronto Press acknowledges the financial assistance to its publishing program of the Canada Council for the Arts and the Ontario Arts Council, an agency of the Government of Ontario.

Canada Council for the Arts   Conseil des Arts du Canada

ONTARIO ARTS COUNCIL
CONSEIL DES ARTS DE L'ONTARIO
an Ontario government agency
un organisme du gouvernement de l'Ontario

Funded by the Government of Canada   Financé par le gouvernement du Canada

Contents

List of Figures vii

Acknowledgments ix

Introduction 3

1 The Law of Similars 27

2 The Law of the Single Remedy 59

3 The Law of Minimum 103

Conclusion 142

Notes 159

Bibliography 185

Index 213

Figures

Note: All illustrations are courtesy of the Institute for the History of Medicine of the Robert Bosch Foundation, Stuttgart.

1 Hahnemann's unpublished repertory of symptoms 39
2 An excerpt from the entry on headache in Hahnemann's repertory 40
3 An entry from Hahnemann's *Krankenjournal* of 1830 73
4 Description of Annette von Droste-Hülshoff from Clemens Bönninghausen, *Homöopathische Heilungs-Versuche*, 1829 80

Acknowledgments

This book would not have been possible without a sabbatical leave from the University of Waterloo. The Lois Claxton Humanities and Social Sciences Research Fund Award at the University of Waterloo helped launch the project, and a Social Sciences and Humanities Research Council Standard Grant allowed me to complete it. An Alexander von Humboldt Research Grant enabled me to deepen my understanding of homeopathy and medicine around 1800 while working at the Institute for the History of Medicine of the Robert Bosch Foundation in Stuttgart. I am indebted to Martin Dinges, Robert Jütte, and their team at the Institute for their quick response to every question and request. Over the months and years, Justus Fetscher has lent invaluable intellectual guidance and assistance, as has his colleague at the University of Mannheim, Thomas Wortmann, with whom discussions about Annette von Droste-Hülshoff and homeopathy have opened up an avenue of future research for me. Several colleagues and students from various institutions have left their stamp on my thinking: I am grateful to the Waterloo Centre for German Studies, the University of Toronto, the University of Pittsburgh, York University, Memorial University, the University of North Carolina at Chapel Hill, Wake Forest University, the University of Pennsylvania, and the Leopoldina in Halle for generous invitations to lecture on Romanticism and homeopathy. Most of all, I cannot begin to express how grateful I am to colleagues in southern Ontario for their interest and encouragement, in particular, Angela Borchert, Christine Lehleiter, and Paola Mayer, loyal and treasured friends at Saturday research lunches. Joan Steigerwald has lent essential, patient guidance in helping me navigate the field of the history of science and medicine. Special thanks go to John H. Smith for our enlivening discussions on Romanticism and for his cheery readiness to read the entire manuscript and offer extensive feedback. Michail Vlasopoulos and Frederick Amrine helped expand my understanding of Spinoza at a recent gathering here in Waterloo. My colleagues at the University of Waterloo have been unfailingly obliging, in particular David John, who answered my every question about Goethe, and Helena Calogeridis, who tracked down innumerable sources for me. Maria Reinhardt did superb work in helping to translate German citations. Finally, I wish to thank my editor Richard Ratzlaff as well as the two anonymous readers from the University of Toronto Press for their insightful, lengthy commentaries.

THE BIRTH OF HOMEOPATHY
OUT OF THE SPIRIT OF ROMANTICISM

Introduction

HOMEOPATHY: A CHILD OF ITS TIME

In *Faust, Part Two* the great German poet, statesman, and scientist Johann Wolfgang von Goethe (1749–1832) made a sardonic reference to a new medical practice of his day – homeopathy. Homeopathy is based on the notion that "like can cure like" or what is known as its Law of Similars. Samuel Hahnemann (1755–1843), the founder of homeopathy, came up with the idea of *similia similibus curentur* in 1796. His idea was that a drug producing symptoms in a healthy person similar to those of an illness could in fact cure this very illness, provided it was given in small enough doses. He published his findings in the main medical journal of the time, *Journal der practischen Arzneykunde und Wundarzneykunst (Journal of Practical Medicine and Chirurgy)*. This journal was edited by Christoph Hufeland (1762–1836), the physician to the Weimar court in the early 1790s and founder of macrobiotics, another natural, alternative medical regimen still popular today.[1] As Hufeland was also Goethe's physician, it is not inconceivable that already in 1796 Goethe was familiar with Hahnemann's article. But by the time Germany's most famous writer finished the second part of *Faust* in 1831, homeopathy had gained in popularity. It also had garnered its share of controversy and ridicule, leading Goethe to let the devil play its impish advocate. An old hag complains to Mephisto that she has an aching foot. He responds by stomping so hard on it that she forgets her original pain: "Zu Gleichem Gleiches, was auch einer litt; / Fuß heilet Fuß, so ist's mit allen Gliedern" ("Like cures like, no matter what one is suffering; / Foot heals foot, and so with every member"; HA 3:195;

6336–7).[2] But the devil also pulls the woman's leg in another way. Misconstruing his allusion to homeopathy, she takes his cruelty as a sexual advance – as playing footsie!

Goethe also mischievously invoked homeopathy in a letter dated 1820, this time referring to another major principle of homeopathy, the Law of Minimum Dose. According to this tenet, the more diluted a homeopathic remedy is, the more efficacious. Goethe had received an amulet with a tiny amount of gold in it. He ironically commented that the Frankfurt jewellers must have heard of Dr Hahnemann's theory and made use of it to their own ends. Goethe here not only calls Hahnemann an amazing, strange ("wundersamen") physician (WA 4.33: 192); he observes that it is equally curious that his controversial teachings have been legitimized and strengthened by their appropriation in a totally different profession – by devious jewellers.

Did Goethe believe in homeopathy? His roguish references to it suggest that he was sceptical of its claims. But two other famous German writers were avid proponents – Bettina von Arnim (1785–1859) and Annette von Droste-Hülshoff (1797–1848). Droste-Hülshoff was treated over a period of years by Hahnemann's student Clemens von Bönninghausen (1785–1864). Von Arnim encouraged the Berlin artist and architect Karl Friedrich Schinkel (1781–1841) in 1838 to try a homeopathic course of treatment, saying that it healed "by gently embracing nature" ("durch sanftes Anschmiegen an die Natur"; Schultz 47). She lauded the excellent and simple effects of homeopathy that sacredly respected nature ("grossartige einfache die Natur heilig schonenndere Einwirkungen"; Schultz 56).[3] Three years later, after years of ill health from a condition that today would be diagnosed as Parkinson's disease, Schinkel died. He had suffered a slight heart attack, after which his condition worsened, causing his physician to prescribe a series of bloodlettings. These eventually led to his death. Goethe, too, at the age of eighty-three, was let two pounds of blood to combat a hemorrhage.

All these luminaries of German nineteenth-century life and letters remind us that the alternative medical practice of homeopathy, so popular today, has a history in the culture in which it arose. In particular, Schinkel's and Goethe's treatments tell us much about the extremes of conventional medical practice. These contrast starkly with Bettina von Arnim's Romantic view of homeopathy as softly and sacredly in tune with a benevolent nature. This historical background, however, is invisible in the consumption of homeopathic preparations today. How many people, when they seek out a qualified homeopath, or on their own go to the local health food store for a remedy, are aware that the product they are taking comes from late-eighteenth-century medical thought? Do they understand the principles behind its effectiveness? Do they know that most of the homeopathic single medications on the shelves were devised by Hahnemann himself?

The purpose of *The Birth of Homeopathy out of the Spirit of Romanticism* will not be to rehearse the two-centuries development of homeopathy from its founding by Samuel Hahnemann to the present day. But it will be to counteract a historical amnesia of a different sort. Rather than offer a diachronistic overview of the expansion of homeopathy from its origins onward, this book ranges widely to look at synchronistic movements occurring at the time of its founding to which it invites comparison. It places Hahnemann's invention of homeopathy within the cultural, medical, and semiotic framework of its day. The basic questions that this book seeks to answer are: Why does homeopathy arise at the time it does? What discursive shifts contribute to its birth and development? Are there earlier models of medicine that paved the way for homeopathy? Are there contemporaneous models or concepts of thought that parallel homeopathy and help explain its growth in popularity? How did Hahnemann conceive the conditions of his knowledge? That is, what is the epistemology that grounds homeopathy? *The Birth of Homeopathy out of the Spirit of Romanticism* gathers together various strands that make homeopathy intelligible as a phenomenon of its time.

The aim of this book is to provide a specific historical framework for educated persons who would browse the homeopathic section in a pharmacy or organic food store and wonder about the little pills, regardless of whether they would buy them, or whether they decide to seek out a homeopath. Homeopaths and naturopaths, too, would find this background pertinent, above all because many of Hahnemann's texts and the scholarship on them is accessible only to those who read German. Given the history of medicine and science I review, as well as the increasing prominence of homeopathy as an alternative health option, physicians could also find this book informative. I presume a reader, though, curious enough about the oddities of late-eighteenth-century medicine to appreciate the detail I provide in order to ensure scholarly integrity and accuracy.

But how did I come to this study myself? Initially, as an amateur herbalist, I was inquisitive about how homeopathy approached many of the herbs I had grown, harvested, and tinctured. I had already known about homeopathy for several years because of its prevalence in Germany, and I have tried its preparations. Then, as a scholar of German Romanticism and upon reading Hahnemann's writings, I began to discern many similarities to ideas with which I was already familiar in my research. The more I investigated the topic, the more clearly I saw that no book in English presented homeopathy as structured by the times in which it emerged. It also became abundantly clear to me what the implications of this scholarly lacuna were. Because, as I hope to demonstrate, homeopathy is a cultural product of its era, it follows that the same cultural prerequisites will never be in place as they were in Hahnemann's lifetime – not 50, not 100, not 200 years later. In short, the practice of homeopathy could not be invented today. But to anyone not familiar with its intricate origins, homeopathy will by default be either considered as evidence of the timeless, inscrutable workings of nature or debunked for being scientifically unverifiable. Why, though, should homeopathy be debated on the grounds of what side offers the more acceptable "proof" when it is a product of historical contingency?

Another way of framing the centrality of historical analysis would be to situate homeopathy within the issues being investigated today by a burgeoning new field, the medical humanities. Johanna Shapiro et al., in "Medical Humanities and Their Discontents," state that this field aims "to improve health care (*praxis*) by influencing its practitioners to refine and complexify their judgments (*phronesis*) in clinical situations, based on a deep and complex understanding (*sophia*) of illness, suffering, personhood, and related issues" (192–3). "Related issues" include, of course, pain, healing, and therapeutic relationships. This article continues by saying that the medical humanities aim to augment the biomedical sciences by encouraging tolerance for

> observations of one's own thinking, emotions, and techniques, recognition of and response to cognitive and emotional biases, and integrating judgment from multiple sources including the scientific, the clinical, and the humanistic. Of special interest is their inclusion of relational, affective, and moral components, including attentiveness, critical curiosity, self-awareness, and presence, dimensions that legitimize introspective, emotional labor as well as instrumental work. (195)

These very concerns have their peculiar legacy within homeopathy and help explain its rise in popularity over the last two hundred years.

To illustrate: if homeopathy advocates today underscore how individual patients time and again attest to their success stories, thereby indicating its effectiveness, it matters that homeopathy appeared at a historical juncture when drastic recourse to opium, emetics, purgatives, and bloodletting were still prevalent. Hahnemann responded to this situation by emphasizing instead the patient's pure recounting of empirical symptoms, requiring an extensive anamnesis and continual self-monitoring and (often epistolary) self-narrative. He will state not that people experience the same illness differently, but that each person has a different illness. Does homeopathy's "truth" or "proof" today rely on the individual's account of its success? Although my book does not examine our complex understanding of these terms in the twenty-first century, terms that render the entire debate about homeopathy's viability polarizing, it does explain what empirical proof meant to Hahnemann and how he conceived of his fidelity to his patients' narratives. An etiology of homeopathy accounts for, among other things, this stress on the individual's experience and narration. In short, if the field of medical humanities "honors rather than dismisses subjectivity" (Shapiro 196), I hope to do the same, though with a decisive twist: I explain how Hahnemann's vision of subjectivity was peculiarly Romantic as well as how "objectivity" was regarded circa 1800. My modest contribution to the field of the medical humanities, then, consists of situating many of its key terms within such a historical framework.

We cannot assume that how we talk about our health remains unchanged from generation to generation. Our ways of perceiving our physical condition as well as conveying our sensations of it rely on the state of medicine and physiology at any given time, culturally specific notions of sensibility, disposition, and pain, as well as shifting dietetic regimens prescribed for the care of the self. Homeopathy, however, has all too often been presented as a singular, isolated invention.[4] Such an approach *brackets* it *outside* rather than *includes* it *within* this history of medicine and the socially prescribed maintenance of body and mind. Yet it is as much indebted to these factors as other medical innovations are. Incontestably, Samuel Hahnemann crafted a new medical practice and merits a prominent place in the history of medicine. Even if he had not invented homeopathy, Hahnemann would have left an impressive legacy. Goethe, for instance, writes that his diet is almost Hahnemannian in its strict limitations, that is, he avoided any food with a medicinal effect, such as spices, coffee, and alcohol (WA 4.40: 276). This pure diet and healthy lifestyle to which Goethe alludes is only one of Hahnemann's insights. In many other respects, he was a radical thinker for his time. In addition to criticizing common harsh medical treatments, he pioneered public hygiene. His records document – more than those of any other physician of the era – copious time spent with patients.

But if Hahnemann is seen exclusively as the great innovator, then his ties to former and contemporaneous models of conceptualizing human health and illness tend to be downplayed. Moreover, if the two centuries–long narrative of homeopathy is constructed with Hahnemann at its start, then the unintended result of this narrative drive to establish clear beginnings is that similarities between Hahnemann and synchronous systems of thought are generally overlooked.

No doubt the chroniclers of homeopathy have been influenced by the rhetoric of its founder. Although a learned, skilled translator fluent in seven languages, Hahnemann promoted himself as an innovator and denied being influenced, except on rare occasions, by medical predecessors. He held his students to a very strict protocol for the examination and treatment of patients, from which he forbade any deviance. He demanded dedicated obedience from his patients and would not let them seek out conventional medical practitioners. He publicly criticized the natural philosophy of Friedrich Wilhelm Joseph Schelling (1775–1856) and his followers among physicians. Hahnemann wrote that "their dualism, their polarization, and representation . . . their potentizing and depotentizing . . . incorporeal and ethereal, soars aloft beyond our solar system, beyond the bounds of the actual" ("On the Value of the Speculative Systems of Medicine," *Lesser Writings* 495).[5] Such vitriolic rhetoric deflects from the fact that Hahnemann himself believed in the incorporeal spirit active in his own potentized remedies. Thus, if Hahnemann consistently presents himself as unique and revolutionary, it requires a special effort to ferret out where his links to his contemporaries lie.

Hahnemann's self-stylization as an individualistic trendsetter who overcame one obstacle after the next in his long career is in many ways itself a Romantic pose. Indeed, the fact that he forged his own persona as a scientific genius and medical prophet encouraged hagiographic deference, as can be seen even to this day in biographies of him. Moreover, being close in age Goethe (1749–1832), he might well have considered his stature comparable to that of Germany's preeminent poet. At the very least, he resembles Goethe's famed Doctor Faust, who similarly despised the deadly doses administered by physicians: the medicines raged, Faust confesses to his assistant Wagner, far worse than the plague they were intended to cure (HA 3:39; 1052). Like Faust, Hahnemann, too, turned his back on inherited traditions passed down by books and looked instead to vital powers. With his desire to find the means to healing through the close observation of nature, Hahnemann seems to echo Faust's longing: "Daß ich erkenne, was die Welt / Im Innersten zusammenhält, / Schau' alle Wirkenskraft und

Samen, / Und tu' nicht mehr in Worten kramen" ("That I may discern whatever / Binds the world's innermost core together, / See all its active forces, and its seeds"; HA 3:20; 382–4). Indeed, Hahnemann could be said to find precisely this "Wirkenskraft" in the dynamized essence of the diluted remedy, the force that resided "Im Innersten." In real life Hahnemann did not compare himself with Goethe or Faust – but he did compare himself with the German reformer Martin Luther ("Necessity of a Regeneration of Medicine," *Lesser Writings* 521).

All in all, we cannot be content with a description of how homeopathy works or of Hahnemann's personality and life. These narratives gravitate to telling an uncomplicated story of homeopathy as being exceptional and timeless. Instead, we want to see what contemporaneous strategies of scientific procedure and proof Hahnemann adopts. How does he fit into prevalent discourses of experiment, observation, and experience? On what basis can scientific evidence be claimed? What constitutes for him authoritative argument? Correspondingly, we want to see how Hahnemann follows and combines narrative and linguistic constructs of the health and illness of his day. How, for instance, does homeopathy relate to other regulatory discourses of the period concerning physiology, dietetics, pathology, morphology, and body–soul interaction? These formations include how an individual owns and narrates his or her pain and what constitutes natural, non-interventionist healing. The linguistic turn in cultural medical studies recognizes that pain, illness, and health depend on how they are expressed, not just in the psychic but also in the somatic arena. What, then, does Hahnemann single out in his reading of the symptoms of illness? If he does not fit them into etiological and prognostic narratives, what representational form do they assume in his records? In other words, how does he mediatize the body in his voluminous writings? Since embodied experience, especially pain, is so difficult to reconstruct, do his patients' discourses match or unsettle medical models of the body? Given the foremost attention Hahnemann paid to the disposition and temperament of his patients, the history of emotions and the self are similarly important factors. Even dialogues about infinitesimal mathematics and animal magnetism play a role in his work. In short, in homeopathy, medicine meets and intersects with a broad cultural field around 1800.

The Laws of Homeopathy

In order to examine this discursive background to which Hahnemann belongs, this book is divided into three chapters, each of which is dedicated to one of the three main principles of classical homeopathy – The Law of Similars, the Law of the Single Remedy, and the Law of Minimum. By focusing on these principles, I hope to convey a sense of the cohesive system or structural model into which Hahnemann re-codes and reorganizes discourses peculiar to the historic moment. I conceptualize homeopathy as a putting into play or enactment of contemporaneous ideas. To a limited extent, I am indebted to the conceptual history of medicine, though not in terms of linear medical progress. To a larger extent, given my training in literary studies, I recognize how medical discourses are semiotic and hermeneutic in nature, that is to say, how symptoms of the body are communicated, categorized, and recorded. My method is to locate in the three major principles patterns borrowed from medical, other scientific, and even philosophical paradigms of the time.

Giorgio Agamben has defined "paradigm" above all as a linking of "singularities." That is to say, rather than establishing causal relations between ideas, thinking in paradigms entails moving from one particular example to the next, often over a span of disciplines. A paradigm forms a heterogeneous ensemble. Following Agamben's cue, I move from Herder (1744–1803), Goethe, Hufeland, Novalis (1772–1801), and Schelling, among others, back to Hahnemann, to investigate what governs homeopathy. Hahnemann's way of thinking led him to striking similarities with these writers, even when a direct lineage cannot be established. In fact, it could be argued that an influence study, showing direct links between Hahnemann and his contemporaries, would be narrow in conception. *The Birth of Homeopathy out of the Spirit of Romanticism* subscribes to a notion of polygenesis, that is to say, ideas arising simultaneously in and across various disciplines.[6]

At times, because Romanticism is a threshold period, a gateway to modernity in terms of thinking differently about the individual subject, health, and nature, older paradigms overlap with emerging ones. Positivistic approaches to the sciences were not yet dominant, and conflicting modes of scientific inquiry vied for prominence. Although Hahnemann in his long life tries to coalesce these discourses into a distinctive constellation, coherent in internal argumentation, nonetheless discrepancies and ambiguities arise. Homeopathy reflects this epochal flux, above all, in Hahnemann's increased reliance on a magical, spiritual energy operative in the remedy. This Romantic view of how homeopathy works belies his origins in medical observation. I draw out such moments when older ideas continue alongside newer ones and where practice conflicts with theory.[7] To put it simply, homeopathy does not suddenly break away from older medical traditions, for it is never the case that one paradigm gives way precipitously to another. Rather, I see homeopathy as archiving multifaceted influences and complex, gradual shifts in medical thinking. Like an archaeologist I want to excavate layered sediments. Again following Agamben, I want to stress heterogeneity and incongruity over uniformity and simplification.

Although I do not argue for Hahnemann establishing a pristine new paradigm, I do wish to claim that homeopathy should be significant to scholars of the late-eighteenth- and early-nineteenth centuries and that Hahnemann deserves to belong to the pantheon of Romantic thinkers. At the very least, a study of Hahnemann offers a more kaleidoscopic view of medical practice and of the vitalistic interpretation of nature circa 1800, topics that not just historians of science but literary scholars of Romanticism are vigorously researching today.[8] But homeopathy is not just one textual construct among others circa 1800 to be investigated by the scholar of Romanticism. It is arguably one of the most important material legacies of that epoch still with us into the twenty-first century, its products found in most pharmacies in Europe and North America. It tells us why the study of Romanticism matters.

What are, though, homeopathy's three principles? And why does Hahnemann make an appeal to principles, in fact, by 1807 to "laws" in the first place, rather than couching his findings as a theory or supposition? It is significant that the originator of homeopathy sought to render it in accord with the laws of nature, on par with how Isaac Newton's (1642–1727) mathematical formulas revealed the laws of mechanics or how Immanuel Kant's (1724–1804) a priori categories were necessary laws of experience. To invoke laws was to make a claim that one's field of inquiry was a science and out of the range of speculation. Specifically for Hahnemann, "laws" would counteract in their neatness the voluminous collections of morbid symptoms he was amassing through his practice. But it is important to recognize that in adopting a rhetoric of "laws" Hahnemann was not unique. In the wake of the scepticism of David Hume (1711–76) and Kantian critical philosophy, a keen awareness existed of the inability to know the inner workings of nature beyond appearances. And so the aspiration grew that the formulation of the laws of mathematics and physics, on the one hand, and critical philosophical reasoning, on the other, would allow one to overcome this inability. Thus, the German physicist Johann Wilhelm Ritter (1776–1810) discovered ultraviolet light while searching for the polarities governing the invisible forces in nature. Novalis, Johann Gottlieb Fichte (1762–1814), and Schelling continued after Kant to critically examine the conditions of human knowledge, even aspiring to the absolute unity of mind and nature because they were cognizant of the limited and incomplete character of what one knew. In medical circles, physicians were similarly preoccupied by defining universal first principles that governed organic life. As Guenther Risse has summarized, "Nature, declared Kant, was capable of being subsumed into a 'pure science' provided it could be based on necessary a priori principles furnished by reason and, most importantly, capable of mathematical expression" ("Kant, Schelling, and a 'Science' of Medicine" 148–9). "Thanks to Kant, medicine could now become a rational 'science' based on laws of general and apodictic character" (149).[9] *The Birth of Homeopathy out of the Spirit of Romanticism* hopes to situate homeopathy within this post-Kantian intellectual framework.

Hahnemann's first principle was laid down in 1796 in his "Versuch über ein neues Prinzip zur Auffindung der Heilkräfte der Arzneisubstanzen" ("Essay on a New Principle for ascertaining the Curative Powers of Drugs"). It was to guide all his subsequent findings. Invoking nature as a touchstone, he writes: "*We should imitate nature*, which sometimes cures a chronic disease by superadding another, *and employ in the* (especially chronic) *disease we wish to cure that medicine that is able to produce another very similar artificial disease*, and the former will be cured; *similia similibus*" ("The Curative Powers of Drugs," *Lesser Writings* 265). Focusing on this early essay by Hahnemann, chapter 1 examines the prevalent medical treatment to which Samuel Hahnemann was reacting and offers examples of *similia similibus curentur*. This Law of Similars illustrates how Hahnemann bridged two different modes of thought. On the one hand, in compiling lists of primary and secondary symptoms produced by substances and matching them with symptoms in his ill patients, Hahnemann belongs to an eighteenth-century paradigm of observational empiricism and data collection. On the other hand, the theory of *similia similibus curentur* is a magical one; it relies on a notion of a suddenly dissimilar, unexpected symptom that the physician singles out in order to select the proper remedy. Hahnemann thus moves from an eighteenth-century architectonics to Romantic speculation in the life sciences and to Romantic theories of analogy and inspired reading.

The second law of homeopathy, the Law of the Single Remedy, one to which Hahnemann strictly held throughout his long career, stipulated that one could not mix remedies. Today at a local health food store, one can purchase several combinations of remedies that will address such general conditions as sleeplessness, allergies, gastro-intestinal discomfort, and so on. Classical homeopathy, however, requires that at any given time a patient be administered only one remedy specific to his or her profile. Hahnemann criticized conventional medicine for attempting to reduce all individual cases to a set of diseases. He insisted that it was always the person with the disease who needed to be treated, not the disease itself.

This focus on the individual has many dimensions. It relates in part to the bedside medicine that Hahnemann professed as opposed to the rise in institutionalized, clinical medical practice. With tensions mounting at the time between the individual bourgeois subject and the specialization of the natural sciences, homeopathy promised non-interventionist, personalized, natural healing. The commitment to each patient can be witnessed in Hahnemann's unique procedure in the consultation room, where he paid special attention to the temperament and affective response of each client. The authenticity of individual response also explains the priority Hahnemann placed on self-testing. Chapter 2, then, examines the hermeneutics of the individual around 1800 as it relates both to Hahnemann's involvement in self-experimentation and to his patient interviews.

Already in the 1796 essay "The Curative Power of Drugs," Hahnemann was reacting against the heroic medicine of his day – the prevalent use of leeches, bloodletting, opium, drastic emetics, and powerful purgatives. By contrast, he noted the effectiveness not just of moderation, but specifically of the small dose. Then in 1799 he announced his principle of the infinitesimal dose, and after 1800, respecting what was to be termed homeopathy's Law of Minimum, he gradually reduced dose sizes. The impact of the catalyst was present even though the toxicity of the substance had disappeared. According to this third law of homeopathy, the homeopathic pharmacological remedy is dynamized by a series of dilutions. Less of the original substance means a more profound effect as a remedy becomes increasingly energetic: the higher the number of dilutions the stronger and deeper the remedy acts. The living spirit within it becomes ever more active. This concept of potentization lies at the heart of German Romantic thought and its search for increased spiritualization.

In chapter 3, then, I look at Hahnemann's own development and understanding of vitalistic principles current at the time. Hahnemann can be seen in line with other thinkers of the day, such as Ritter, Friedrich Wilhelm von Schelling, and Lorenz Oken (1779–1851), who imparted properties of life to nonliving matter. Mesmerism, for instance, a form of treatment that Hahnemann included in his practice, is indebted to this notion of an invisible, dynamic force present in nature which can be manipulated in individual patients.

By 1807–8 Hahnemann was calling the principle of *similia similibus curentur* a law of nature. The Conclusion is devoted to the belief in a spiritualized nature, in harmony with man, as expressed by such Romantics as Novalis, Schelling, and Oken, and as influenced by the philosopher Baruch Spinoza (1632–77).

Hahnemann's Life and Work: A Synopsis

But who was Samuel Hahnemann and what are his most significant publications?[10] The father of homeopathy was born in 1755 in Meissen, son of a painter to the famous porcelain manufacturer. He studied medicine in Leipzig and in Vienna under Joseph von Quarin (1733–1814). Unlike in Leipzig, which had no hospital or clinic of its own for the practical training of medical students, the Hospital of the Merciful Brothers, which Quarin directed, had ample hospital beds where the young student could collect experience. Another prominent Viennese doctor, Anton Störck (1731–1803), might well have served as Hahnemann's inspiration for the self-testing of drugs. From Vienna in 1777, Hahnemann accompanied the Baron von Brukenthal to Hermannstadt, Transylvania, where he served as his private physician, librarian, and superintendent of his coin collection. At this time he became initiated in the Masonic lodge, a not insignificant fact in his life, given his life-long optimism and commitment to the betterment of mankind, values that are prevalent during the Enlightenment but especially pronounced in Freemasonry. These values help explain the deistically inflected references in Hahnemann's writings to God as the "Creator" and "Instructor of Mankind" ("The Medicine of Experience," *Lesser Writings* 436–7), the "great Spirit of the Universe" ("The Medicine of Experience," *Lesser Writings* 440), and to a beneficent nature ("Old and New Systems of Medicine," *Lesser Writings* 723).

In 1779 Hahnemann was granted the doctoral degree in medicine in Erlangen, after which he pursued an itinerant medical career until settling down for a longer period in Torgau in 1805. During this early period in his medical career he published several popular works and pamphlets, including *Freund der Gesundheit* (*The Friend of Health*) (1792 and 1795). Many of his shorter writings were published in popular journals, not in venues for specialists, demonstrating his belief in educating the public. Indeed, later he would stipulate that his *Organon* was compulsory reading for his clientele, involving them in their own self-awareness and knowledge of homeopathy. Before the turn of the century Hahnemann had also translated numerous writings as diverse as Paul-Henri Thiry d'Holbach's *Système de la nature*, Joseph Berrington's 638-page *History of the Lives of Abelard and Heloisa*, Jean-Jacques Rousseau's *Handbook for Mothers or Principles on the Education of Infants*, and William Cullen's *Materia medica*. By 1806 he finished translating Albrecht von Haller's *Materia medica*, which discussed over four hundred native herbs.

In the translation of Cullen of 1790, Hahnemann for the first time published his unreserved criticism of the pernicious treatments of bloodletting, emetics, and purgatives. Already at this initial stage in his career, then, he became known for a sensible, restrained approach to healing. In the *Freund der Gesundheit*, for instance, he warns about visiting the sick in order to prevent contagion ("The Friend of Health, Part I," *Lesser Writings* 164), the need for airing sick rooms, and how candles spoil the air (177). Seeing each patient as unique, he advised against generalized dietetic rules (188). He addressed the necessity of care for the ill in prisons (215), public schools (226), and orphan asylums (227), and warned of pestilence in military hospitals (216) and on ships (219). Even in an era before Rudolf Carl Virchow's (1821–1902) germ theory of disease and the ensuing drills of quarantine, disinfection, and sterilization, he counselled against kissing others, handshaking, drinking from their glass, and using their toilet (217). In fact, later, in response to the cholera epidemic of the early 1830s he wrote of the "brood of ... excessively minute, invisible, living creatures" ("The Mode of Propagation of the Asiatic Cholera," *Lesser Writings* 758).

The last decade of the eighteenth century saw two significant developments in Hahnemann's career. First, with the support of the Duke Ernst von Sachsen-Gotha, Hahnemann established a convalescent home for the mentally ill in Georgenthal. Although he was unsuccessful in bringing other patients to the home, he did treat one individual there, the writer and secret chancellery secretary Klockenbring.[11] Like Philippe Pinel (1775–1826), who was freeing lunatics from their chains in Paris at this time, Hahnemann prohibited physical punishment and restraint. This considerate handling presages the prime importance he allotted later in his practice to the emotional state in the anamnesis, and his recognition of the need to create bonds of confidence between physician and patient. The other salient development of the 1790s was Hahnemann's self-experimentation with Peruvian bark, otherwise known as cinchona and known as a remedy against malaria. He noted that taking cinchona produced in him symptoms of an intermittent fever similar to those produced in malaria, leading him eventually to expound the prime law of homeopathy, *similia similibus curentur*.

In relation to the development of a specific terminology, Hahnemann began to deploy the word "dynamic" to characterize the healing process as early as 1797.[12] By 1807, by which point he had extracted fifty remedies, Hahnemann defined the adjective "homeopathic,"[13] and first used it as a noun in 1810 in the *Organon der rationellen Heilkunde*.[14] The word *homeopathy* stems from the Greek *homoios*, meaning similar, and *pathos*, meaning sickness or feeling. The *Organon*, Hahnemann's major treatise, underwent six versions from 1810 to 1842, although the last edition only saw print in 1921 because his widow withheld permission to publish it while she was alive.[15] The term *rationell* designates the scientific systematization or abstraction of the laws of nature that Hahnemann formulates. By the second edition, Hahnemann dropped this word and changed the title from *Heilkunde* to *Heilkunst* in order to indicate that his treatise was not theoretical in nature (*Kunde*) as much as dedicated to the practical art (*Kunst*) of healing.[16] Still, the *Organon* refers to the homeopathic *laws* of nature and healing ("Naturgesetz" [§53] and "Heilgesetz" [§178]), whereby Hahnemann defines health not as relative, as in the Brunonian model, and consequently healing not as the mere therapeutic lessening of symptoms. Instead, the cure is long-lasting and permanent. The first two paragraphs of the *Organon* read: "§1. The first and *sole* duty of the physician is, to restore health to the sick. This is the true art of healing. §2. The perfection of a cure consists in restoring health in a prompt, mild, and permanent manner; in removing and annihilating disease by the shortest, safest, and most certain means, upon principles that are at once plain and intelligible" (95). The other major word that Hahnemann coined still in currency today is *allopathy*, which he used in the 1816 preface to the first edition of volume 2 of the *Reine Arzneimittellehre* (*Materia Medica Pura*). Whereas the homeopathic preparation produces an illness similar to the one it is designed to reverse, allopathic medicine elicits a response contrary to the illness it intends to cure.

Alongside the *Organon*, which underwent translation into many European languages during its author's lifetime,[17] Hahnemann's other major publications collect the so-called drug provings or testings. These *materia medica* are the alphabetical listing of medicines and their effect on the human body. But unlike traditional pharmacopoeia which list the symptoms of the sick that the medicine is intended to cure, the homeopathic version lists the effects of a substance as tested on the healthy human body (which only subsequently are matched to the symptoms of the patient). Hahnemann's first homeopathic pharmacopoeia was the *Fragmenta de viribus medicamentorum* (1805) in which twenty-seven remedies are listed,[18] forerunner to the German-language six-volume *Reine Arzneimittellehre* (first edition 1811–21; second edition 1824–7), listing sixty-three remedies. The term *reine* or, in the English translation, *pura*, refers to the pure symptoms of a drug, that is, as produced on a healthy individual. One of the major differences between the *Fragmenta* and the *Reine Arzneimittellehre* was that in the latter Hahnemann listed the symptoms from head to toe rather than haphazardly. Then, in his last major work, the five volumes (beginning in 1828 with volume 1) of *Die chronischen Krankheiten* (*The Chronic Diseases*), likewise a compendium of remedies, the emotional symptoms of a patient are listed first, indicating their centrality.[19] It should be also noted that Hahnemann possessed, in addition to these *materia medica*, an unpublished repository that he frequently consulted. This collection alphabetically listed not plant remedies but symptoms. Under each major symptom, Hahnemann pasted tiny slivers of paper, each devoted to minutely describing varying indicators. Each of these subentries then notated the matching remedy. For instance, there are over thirty pages of subentries under the rubric headache.

Following his seven-year stay in Torgau, from 1811 to 1821 Hahnemann lived in Leipzig, where his practice, teaching, and popularity grew. But his principles of healing ran him afoul of the apothecaries, who eventually drove him out of town and to Koethen, where he lived between 1821 and 1835. On various issues, Hahnemann's recommendations were contrary to common practice. He advocated simple medication, not a poly-pharmacy of concoction. In other words, he was against prescribing several remedies at once for various symptoms ("Are the Obstacles to Certainty and Simplicity in Practical Medicine Insurmountable?" *Lesser Writings* 319; "A Preface," *Lesser Writings* 345, 348; "Cure and Prevention of Scarlet-fever," *Lesser Writings* 374), noting that medications can act in opposition to one another ("The Medicine of Experience," *Lesser Writings* 470). By the same token, he also wrote that physicians falsely lumped diseases together in order to treat "by the same medication, with a small outlay of trouble!" ("The Medicine of Experience," *Lesser Writings* 443). Aware of the toxicity of drugs, and reacting against what Michel Foucault later termed the "Birth of the Clinic," Hahnemann noted the peril for patients of experiments and trials in hospitals. Finally, he relied on his own preparation of drugs. It was this latter stipulation against which the Leipzig pharmacists protested, forcing Hahnemann to close his Leipzig practice.

Clearly, they were eager to get rid of him: Hahnemann harshly criticized dispensaries for their exclusive, monopolizing sale of drugs ("Aesculapius in the Balance," *Lesser Writings* 429). He noted that the "whole system [was] . . . for the benefit of . . . the profit of apothecaries" (ibid. 421). He lambasted the lavish waste of costly drugs and even the expensive after-treatment arising from the ill symptoms they produced ("On the Value of the Speculative Systems of Medicine," *Lesser Writings* 488; "Necessity of a Regeneration of Medicine," *Lesser Writings* 519). He came to the conclusion that diseases can be caused by medicines themselves ("Necessity of a Regeneration of Medicine," *Lesser Writings* 418), especially chronic diseases (*Organon* 138, §74). Allopathic medicines also impeded the recovery of the body (*Organon* 175, §156; and 190–1, §207). Furthermore, he noted how physicians conceal disagreeable effects of medications from their patients ("The Curative Powers of Drugs," *Lesser Writings* 297) and how they find it convenient to dispense medications for depression rather than attack its causes ("Are the Obstacles to Certainty and Simplicity in Practical Medicine Insurmountable?" *Lesser Writings* 317). They avoided treating chronic diseases ("Aesculapius in the Balance," *Lesser Writings* 414). The doctor was a "mechanical workman" who merely "writes prescriptions . . . for whose effect he is not answerable" (ibid. 433). This so-called expert only needed to remember a few prescriptions ("Old and New Systems of Medicine," *Lesser Writings* 714) and ignored diverse symptoms and absence of others in making a diagnosis (ibid.), all while never lacking "plans of treatment . . . as long as [the patient's] purse, his patience, or his life lasted" (ibid. 716). Hahnemann parodied physicians who would increase dosage, just because a medication was not working ("Allopathy: A Word of Warning to All Sick Persons," *Lesser Writings* 747). As to their general disposition, he noted their superciliousness ("Three Current Methods of Treatment," *Lesser Writings* 528, 530). By contrast he saw the need of a "sympathizing and attentive physician" ("Old and

New Systems of Medicine," *Lesser Writings* 718) and regarded the profession as a "matter of conscience" (ibid. 723). He insisted that homeopaths be dedicated to treating the poor ("Allopathy: A Word of Warning to All Sick Persons," *Lesser Writings* 751). From its inception, then, homeopathy was conceived in opposition to the sovereignty, institutionalization, and profit making of the medical system, all of which went hand in hand with the latter's disciplined mass control of patients.

The final, stunning development in Hahnemann's personal life merits mention. On 7 October 1834, a young but wealthy patient, Mélanie d'Hervilly (1800–78), travelled from Paris to Koethen to be treated by the famous doctor. At the time Hahnemann was eighty years old and had been widowed since 1830. Despite their age differences, the two fell in love, were married, and moved to Paris, where Hahnemann, with his young wife by his side, continued to practise with a sizeable, wealthy clientele. In fact, with the rich and famous, including Paganini, now seeking out the newest, fashionable medicinal treatment, homeopathy gained a popularity not seen to the same extent in Germany, facilitating its expansion across the globe. Even illustrious men of the time, such as Lord Elgin, came to pay respects to the founder of homeopathy. Hahnemann, who even during his Paris years maintained a relatively youthful appearance, passed away in 1843, at the ripe old age of eighty-eight.

Theory versus Praxis?

As I mentioned earlier, given Hahnemann's long life and medical praxis, it becomes necessary to recognize the various, often contradictory moments in his thought and its development. The scholar of Hahnemann can aim to synthesize his thought, but it is imperative as well to acknowledge discrepancies that arise from the *Gleichzeitigkeit des Ungleichzeitigen* (simultaneity of the non-simultaneous), which is to say, the synchronicity of overlapping, contradictory discursive paradigms. For instance, given that he was born in 1755, Hahnemann's formative years lie within the German Enlightenment and much of the scholarship on him places him strictly, perhaps all too exclusively, within eighteenth-century rationalist and deistic beliefs.[20] He is seen as an empirically minded scientist who relied exclusively on the observation and recording of data. But Hahnemann came up with his notion of *similia similibus curentur* in 1796, when he was already forty-one years old, and coined the term homeopathy at the mature age of fifty-five. In the history of German life and letters, by 1796 the Age of Enlightenment had been superseded by Classicism and Romanticism. At this time Schiller (1759–1805) and Goethe had their most productive intellectual exchange.[21] The Jena Romanticists, Novalis and Friedrich Schlegel (1772–1829), had just begun to publish significant writings. Again, to invoke the *Gleichzeitigkeit des Ungleichzeitigen*, both Classicism and Romanticism were living side by side.

As the years passed and Hahnemann refined his concepts of homeopathy, he turned to increasingly speculative notions that resemble more Romantic than empirical tenets. For instance, starting in the third edition of the *Organon* he recommended increasingly higher potencies, all the way up in the sixth edition to the so-called Q-potency or the LM potencies, in which the active ingredient, though hardly material, was believed to be all the more spiritualized. As well, he referred more and more to a vital life force (*Lebenskraft*) within the body: although invisible, it underpinned the organic road to healing. This belief in invisible but omnipresent forces bring him into line more with Romantic speculation and its predilection for the vocabulary of *Potenzierung* than with Enlightenment rationalism; which is not to say, however, that Hahnemann did not continue throughout his life with minute collection of data based on empirical observation. Hahnemann considered himself as working inductively. He did not, like Schelling, set out first to propose and then deductively validate a theory of the unity of nature and spirit. Yet his writings led to the same conclusion as the *Naturphilosophen* in their common belief in a vitalism that permeated organic and inorganic matter.

Another way of expressing these tensions (as will be delved into more in chapter 3) would be to say that Hahnemann was Kantian in his belief in an epistemology based on empirical observations, yet post-Kantian in his insistence on the actuality of *Lebenskraft*, which Kant believed was only a postulate and could not be proved. For Hahnemann, although one could not manifestly see the vital core or life force in nature (*natura naturans*), one experiences nature (*natura naturata*) as a posteriori evidence of this force: immaterial spirit was hence concretized. Goethe and Alexander von Humboldt (1769–1859), too, confronted such conundrums. First, like Hahnemann, they were staunch empiricists, yet they also demanded to see into the life of things. Second, although they were dedicated to vibrant empirical perception and against the philosophical and theological system building of his time, none of them separated objective observations from their subjective registering.

Other paradoxes surface to make Hahnemann a fascinating yet controversial figure. He bridged external symptoms and internal energy, linking materiality with the intelligibility of life. He camouflaged his subjectivity with a language of technical precision. His *Weltanschauung* was both spiritualistic and naturalistic. Although Hahnemann grounded his belief in the healing power of drugs in a deistic view of creation, he also saw nature as autogenetic and searched for her profound, underlying principles. No less than the famous novelist and aesthetician Jean Paul Richter (1763–1825) pondered the discrepancies in Hahnemann's character. Noting the physician's rare vision as well as his careful studiousness, Jean Paul observed that Hahnemann represented "an odd Janus head [Doppelkopf] of philosophy and erudition" (292). First and foremost, however, the most salient paradoxes of homeopathy are that (1) the smaller the dose the more powerful the remedy is; (2) the toxicity of medicines can thereby be curtailed and rendered curative; and (3) Hahnemann's theories resist being proved wrong, yet they also resist understanding.[22]

Not surprisingly, what I have called here tensions and paradoxes also pervade the scholarship on Hahnemann. Here I wish to lay my cards on the table in an ideological playing field where, given homeopathy's prominence worldwide as an alternative medical practice, not to mention its marketing presence, the stakes are very high. My focus on the three major laws of homeopathy mean that from the start I perceive Samuel Hahnemann as a systematic thinker, that is to say, someone who organizes and crystallizes his findings in terms of principles. I follow here the lead of the pre-eminent German historian of medicine Karl Rothschuh, who termed Hahnemann a "systems thinker" (*Konzepte* 336) and who referred to the "logic of the homeopathic system" (340). Thus, despite inconsistencies in his findings, indeed, in a ploy to overcome them, Hahnemann, I would argue, attempts to build a coherent *Denkstruktur*.

By *Denkstruktur* I mean that the doctrines of homeopathy aim to coalesce into a tight system. The particulars of the provings and patient records, although based on empirical observation, serve to support the pre-existing structures or principles. To give a salient example, the *Reine Arzneimittellehre* is a compilation *not* of what remedies have been proved to work in curing the ill *but* of what symptoms a substance causes when tested on a healthy person. Here it must not be forgotten that Hahnemann tested first and foremost on himself. Hahnemann's justification for this counter-intuition is his law of *similia similibus curentur*: it only makes sense to collect and table symptoms of what a substance causes in the healthy, if it is maintained *in principle* that this very same substance, administered in minuscule dosage, will cure the same symptoms in the sick. Put simply, if Hahnemann has already decided according to his laws the outcome, then it doesn't matter what the patient's reactions are. Instead, the production of symptoms (primarily Hahnemann's own) listed in the *Reine Arzneimittellehre* forecast the result. Whatever new symptoms a patient might develop merely occasion a switching of remedies.

Throughout the extensive literature on homeopathy, however, the tendency has been to emphasize Hahnemann as a medical practitioner and not as a "systems thinker." This accentuation pits his practice against his theory, favouring the former. The reason for this privileging is clear: to perceive homeopathy as based on empirical experience verifies its validity. To intimate that it is a *Denkstruktur* seems to question its legitimacy.

Such a debate has a history as long as that of homeopathy. Hahnemann saw himself as belonging to the Hippocratic tradition of medical experience. By contrast, he sharply attacked the apothecaries and physicians of his day for their blindness to the practical, therapeutic implications of the medicines they were dispensing. This antagonism between theory and therapeutics heightened in the course of the nineteenth century. Already in 1826, a certain D. Rummel, in describing the differences between homeopathy and Brunonian medicine (that Schelling so enthusiastically received), claims that John Brown (1735–88) had erected a system, whereas for Hahnemann theory is of secondary importance: the homeopath relies on sober observation of nature (5). Hahnemann's student Constantine Hering (1800–80) then makes the startling statement in his preface to the 1836 English version of the *Organon*: "I have never yet accepted a single theory in the Organon as it is there promulgated . . . It is the genuine Hahnemannian spirit totally to disregard all theories, even those of one's own fabrication, when they are in opposition to the results of pure experience" (qtd. in Treuherz, *Genius of Homeopathy* 75). In 1882, Hahnemann's translator, Robert Ellis Dugeon, makes the sweeping claim, rejecting the "treacherous quicksands of conjecture" (qtd. ibid. 179) that dominated Hahnemann's post-Leipzig career: "Hahnemann's is the one name in the whole history of medicine connected with a rational, simple and efficacious system of therapeutics, based on the solid foundation of impregnable facts" (qtd. ibid. 188). This dividing line between theory and empiricism continues today. To give a few examples: Harris Coulter aligns Hahnemann with an empirical medical tradition, as opposed to what he terms methodism. And Schmidt insists that what makes homeopathy therapeutically relevant today is a successful praxis not theory ("das Ähnlichkeitsprinzip" 171–2).

In recent years, the impressively annotated new German editions of Hahnemann's writings have sustained the focus on Hahnemann's experience and praxis. Various editors even pointedly downplay homeopathy as a system. In his introduction to *Die chronischen Krankheiten*, for instance, Will Klunker maintains that Hahnemann kept his distance from the influence of hypothetical assumptions (xvi) and that homeopathy, like chemistry and physics, is based entirely on experience (ix). To underscore his point, Klunker observes that the ratio of theory to praxis in *Die chronischen Krankheiten* is 1:20, although this ratio would be expected in a compendium of symptoms elicited by botanical and mineral substances. More nuanced is the remark of Christian Lucae and Matthias Wischner in their introduction to the *Gesamte Arzneimittellehre*. They say that, without Hahnemann's testings of substances and their publication, there would not have been homeopathy as a practice; to be sure, homeopathy would have existed as a theory on paper, with the *Organon* as its foundation. But, they maintain, homeopathy would have disappeared like so many of the theoretical medical concepts of the day did it not prove suitable for application (8).

Like the new editions of the *Reine Arzneimittellehre* and *Die chronischen Krankheiten*, the recently published transcriptions of and commentaries on the *Krankenjournale* (Hahnemann's day-by-day entries about his patients' visits) likewise focus on compilations that result from the physician's empirically based observations. The *Krankenjournale* can be seen as parallel data collections to the *materia medica* (published alphabetical listing of remedies) and to the repository (unpublished alphabetical listing of symptoms), though with focus on the patients' symptoms not the testers'. Like the *materia medica* and the repository, the *Krankenjournale* testify to the prodigious notation of symptoms that Hahnemann conducted throughout his lifetime.[23] To introduce order into such voluminous recording of patient symptoms, the editors of the *Krankenjournale* (Bußmann, Fischbach-Sabel, et al.) gather statistics on the age and sex of Hahnemann's patients, the remedies prescribed, the types of symptoms recorded, and so forth. But they also interrogate how Hahnemann's praxis compares with his principles and whether it holds up to them. For instance, the question arises whether Hahnemann dispenses remedies variedly – based, as he stipulated, on the uniqueness of each case – or whether, in a given time span, he prescribed the same remedy in case after case.

In addition to this extraordinarily painstaking and meticulous editorial scholarship being conducted in Germany today, the published research by social historians of homeopathy, most notably by Martin Dinges and Robert Jütte, also concentrates on the day-to-day praxis of Hahnemann and his students. Here too, despite the prevalence of homeopathy outside German-speaking lands, the language barrier limits the wide readership their important contributions deserve. Jütte is the director and Dinges the deputy director and archivist of the Robert Bosch Foundation's Institute for the History of Medicine (Institut für Geschichte der Medizin, or IGM), which houses Hahnemann's correspondence and the *Krankenjournale*. This archive is sizeable, with 54 *Krankenjournale* from 1801 to 1843 and roughly 5500 letters from 1800 to 1843. It also maintains an unrivalled library of scholarship on the history of homeopathy.[24] In contrast to the bibliographical scholarship that portrays Hahnemann as a genius, that is to say scholarship that belongs to the genre of medical history from the top down, the studies coming from the IGM (Baschin; Bleul; Busche; Dinges; Faure; Genneper; Heinz; Hickmann; Papsch; Plate; Schreiber) examine medical history from the perspective of the lives of common folk, that is, from the ground up.[25] It is an *Alltagsgeschichte* that focuses less on the canonical writings of Hahnemann than on the non-canonized work. In addition to the *Krankenjournale*, scholars at the IGM examine the archive of letters, testimonies, biography of patients, demographics (noting differences of sex, age, and class), patient education, medical lay associations, and tradition of homeopathic self-medication.[26] They attend specifically to the range of illnesses, the cost, and medical topography of urban versus village. The correspondence with Hahnemann offers a particularly unique source because Hahnemann's practice was so distinctive that many patients could not afford to travel long distances to him and conducted their treatment instead by letter. Another salient topic at the IGM is the role played by medical institutions and professionalization, specifically, the spread of homeopathy across continents in the last 150 years. Research affiliated with the IGM thus moves away from studying homeopathy as a conceptual theory to focus on the patient–physician interaction and its development as an institution.

Barbara Duden has described her work in *Alltagsgeschichte* as instrumentalizing "beliefs about the body that are guided by medical praxis" (206) and states that she is not interested in the "ideology of the physician" (207).[27] The same circumspection governs the writings coming recently from the IGM. The drawback of such judiciousness and caution is that it glosses over many aspects of Hahnemann's "beliefs about the body" that are not necessarily driven by his praxis. But even medical praxis is determined by contemporaneous discourses and technologies of the body and is not untainted by them; praxis is not somehow free from "ideology." Unfortunately, Duden's term "ideology of the physician" is dismissive and misleading: it leads to the fear that, if the scholar focuses too intently on Hahnemann's theories about the body and mind, the founder of homeopathy might be written off as an ideologue. It is a risky sort of trepidation, because it will bypass studying homeopathy as a hypothetical *Denkstruktur* governed as much by historical specificity as by medical praxis. For instance, given all the confusing, varying strains of medicine as well as discoveries in the fields of chemistry, physics, and physiology in the eighteenth century and at the start of the nineteenth, it is no wonder that Hahnemann would try to devise a coherent, direct scheme or theory – and one he clearly laid down in the *Organon der Heilkunst* – to explain illness and its cure. I do not deny that Hahnemann was empirically dedicated; but I do insist that notions of law, empiricism, and objectivity in experimentation are themselves historically dependent and differ from our contemporary expectations in medical science and technology.

Is homeopathy fact or fiction? By now it should be clear that I find such a question reductive and facile. I do not set out in this book to evaluate whether homeopathy is a viable medical treatment today. I do not try to explain in some contemporary fashion how homeopathy might be plausible or how its remedies might produce positive results. If I do not engage in the question of why homeopathy might work for either physio-chemical reasons or because of the placebo effect, it is because such questions are out of line with my project. I am not a medical practitioner but an intellectual historian. Such a circumspect stance is the most honest and constructive one for me to take.

But I do leave it up to each reader to draw his or her own conclusions about homeopathy after reading this book. While I remain at the level of studying homeopathy as a historical phenomenon, I believe that history matters for the present. It is crucial to understand that commercialized homeopathic remedies today differ in few respects from how and to what purpose Hahnemann created them two hundred years ago. Because of this similarity, the historical strands informing homeopathy's origins should matter to anyone voicing an assessment of it. The fact that different modes of scientific inquiry, especially in the medical field, ended up being victorious does not mean that we should not look at this earlier period. What has happened, though, is quite the opposite: a bias towards the present has led, on the one hand, the scientific community to interrogate homeopathy in terms of evidence-based medicine and, on the other, its practitioners to swear by its effectiveness. I find it far more productive and reasonable to shift the debate onto a different playing field entirely and explicate homeopathy as a child of its time, *precisely because it can be made intelligible via its historical context*. "How can a highly diluted substance heal?" In *The Birth of Homeopathy out of the Spirit of Romanticism* I hope to provide a historical explanation of this and many other conundrums presented by this prominent but contested alternative medical modality.

Chapter One

THE LAW OF SIMILARS

Similes: Medicinal and Literary Comparisons

The best entry into the world of homeopathy is via Samuel Hahnemann's own inaugural essay of 1796, "Versuch über ein neues Prinzip zur Auffindung der Heilkräfte der Arzneisubstanzen" ("Essay on a New Principle for Ascertaining the Curative Powers of Drugs"). Here he first vocalized his principle of *similia similibus curentur*, which was to guide all his subsequent findings: "*We should imitate nature*, which sometimes cures a chronic disease by superadding another, *and employ in the* (especially chronic) *disease we wish to cure that medicine that is able to produce another very similar artificial disease*, and the former will be cured; *similia similibus*" ("The Curative Powers of Drugs," *Lesser Writings* 265).[1] Hahnemann arrived at the principle of *similia similibus* via his contention that the conventional medicine of his day operated via the principle *contraria contrariis*, which stipulated an illness should be treated with a drug producing the opposite effect. Hahnemann thus begins this essay by rehearsing the problems of medicine of his day, a rhetorical strategy that he similarly adopts in the various editions of the *Organon der Heilkunst*. Other important essays that also take contemporaneous medical practices to task include his "Fragmentarische Bemerkungen zu Browns *Elements of Medicine*" ("Fragmentary Observations on Brown's *Elements of Medicine*," 1801), "Monita über die drey gangbaren Kurarten" ("Three Current Methods of Treatment," 1801), "Aeskulap auf der Wagschale" ("Aesculapius in the Balance,"

1805), and "Ueber den Werth der speculativen Arzneysysteme" ("On the Value of Speculative Systems of Medicine," 1808).

Clear examples of *contraria contrariis* would be counteracting constipation with purgatives, pain with opium, or acidity in the stomach with alkalis ("The Curative Powers of Drugs," *Lesser Writings* 261). To comprehend the significant difference that Hahnemann offered with his concept of the minimal dose and its appeal to patients who were otherwise at the mercy of radical treatments (although today it is, for better or worse, hardly considered radical to treat stomach acidity with alkalis, such as Tums), it is important to understand medical practice around 1800. The notion of offsetting or neutralizing the cause or source of an illness was prevalent because of widespread belief that one needed to "expel from the body that imaginary and supposed material cause of disease" ("Introduction," *Organon* 29). This expelling took the form not only of diuretics, emetics, and purgatives, but also of bloodletting, which was commonly prescribed for various manifestations of inflammation. Hahnemann writes: "They recommend diaphoretics, diuretics, venesection, sectons, and cauteries, and above all, excite irritation of the alimentary canal, so as to produce evacuations from above, and more especially from below, all of which were irritatives" ("Introduction," *Organon* 41). He refers to "the old school of medicine . . . [that] still imagined they could arrest disease by a *removal of the supposed morbid material cause*" (ibid. 29). He even mentions the incidence of a "young girl, of Glasgow, eight years of age, having been bitten by a mad dog, the surgeon *immediately cut out the* part, which, nevertheless, did not save the child from an attack of hydrophobia" (ibid. 36). Hahnemann was deeply opposed to such drastic treatments that intended to expel, excise, or otherwise remove the cause of illness. In fact, in 1808 he claimed that for "twelve years I have used no purgatives . . ., no cooling drinks, no so-called solvents or deobstruents, no general antispasmodics, sedatives, or narcotics, . . . no general diuretics or diaphoretics, . . . no leeches or cupping glasses," and so forth ("Necessity of a Regeneration of Medicine," *Lesser Writings* 517).

In addition, in the "Versuch über ein neues Prinzip" Hahnemann objected to what he termed the "paltry modes" of ascertaining the powers of medicines, which then bore "the stamp of their worthlessness" ("The Curative Powers of Drugs," *Lesser Writings* 253). He protested the testing on animals with the exclamation: "How greatly do their bodies differ from ours!" (ibid. 253). He baulked at the chemical testing in vials, as if the fluids in the body acted the same (ibid. 252). The external signs of plants, as in the Paracelsean teaching of signatures (ibid. 254), were not an indication of their curative powers, nor were their botanical affinities, because plants belonging to the same family could have different effects (ibid. 255). He also objected to empirical trials in hospitals, for they were guided not by therapeutic goals but by scientific principles and were conducted at the peril of the patient (ibid. 257). Opposed to speculative theories of disease origin, he in particular challenged nosologists who could not separate "the essential from the accidental" (ibid. 260): one could not ascertain and then remove the fundamental cause of a disease if it remained concealed (ibid. 261).

Above all, medication that aimed at producing the opposite condition was not only merely palliative but, in fact, injurious and destructive precisely because of its temporary nature, which could result in the aggravation of the original condition when the latter returned. After a brief period of apparent relief, the original illness would break forth again. Moreover, "the disease plants its roots still deeper" ("The Curative Powers of Drugs," *Lesser Writings* 262). The reason the disease returned more grievous than before, Hahnemann argued later in the *Organon* with respect to laxatives, was that "the ill-advised evacuations have lessened the energy of the vital powers" ("Introduction," 45). Another problem with palliatives was that, although they could suppress an original malady, a new disease could appear as a result of their consumption ("The Medicine of Experience," *Lesser Writings* 457). For instance, opium could be taken for sleeplessness, but it could cause a host of other problems. In the essay of 1796 he offered as a clear example of the rebound effect in the case of opium: at first it induces a "fearless elevation of spirit, a sensation of strength and high courage, an imaginative gaiety," only to be followed by "dejection, diffidence, peevishness, loss of memory, discomfort, fear" ("The Curative Powers of Drugs," *Lesser Writings* 266). He perceptively noted: "Chronic pains of all kinds are still sought to be removed by the continued use of opium; but again, with what sad results" (ibid. 262). In fact, the reason "palliative remedies do so much harm in chronic diseases, and render them more obstinate" is because the secondary reaction is "similar to the disease itself" (ibid. 267). All remedies have these biphasic effects. Indeed, if a remedy, he sardonically observes, is claimed not to have the slightest bad effect and yet supposedly cures the worst diseases, we know that it is perfectly ineffective (ibid. 297). Later he was to write that if a disease seems, so to speak, cured by conventional medication, it is only by chance, occurring while the body has become preoccupied by the new disease the medicine occasions ("Aesculapius in the Balance," *Lesser Writings* 418).

It was this secondary, indirect action, following upon the antagonistic, direct action that led Hahnemann to conceive of the notion of *similia similibus*. If a drug could be administered in small doses, it could produce the counter-effect of the strong dose: for example, "valerian (*valeriana officinalis*) in *moderate* doses cures chronic diseases with excess of irritability, since in large doses . . . it can exalt so remarkably the irritability of the whole system" ("The Curative Powers of Drugs," *Lesser Writings* 269). Hahnemann is very clear in stipulating that *similia similibus curentur* does not mean "assisting and promoting the efforts of nature and the natural crisis" ("On the Present Want of Foreign Medicines," *Lesser Writings* 487).[2] For instance, it was commonly argued that by using emetics one aids the natural bodily reaction of vomiting, or by using laxatives one enhances the normal process of defecation. But for Hahnemann, these remedies that could mimic bodily functions were extremely injurious, although they were conceived at the time, still adhering to the Galenic model of abetting flows in the body, to be thoroughly natural, following the doctrine of *vis medicatrix naturae*.

Another example of *similia similibus curentur*, among the many offered in the 1796 essay,[3] is coffee, which can produce headaches in large doses but can cure them in smaller doses ("The Curative Powers of Drugs," *Lesser Writings* 271–2). Hahnemann adds that "other abnormal effects it occasions might be employed against similar affections of the human body, were we not in the habit of misusing it" (ibid. 272). He was to later write in the *Organon*: "Strong coffee in the first instance stimulates the faculties (primitive effect), but it leaves behind a sensation of heaviness and drowsiness (secondary effect), which continues a long time if we do not again have recourse to the same liquid (palliative)" (131, §65). Coffee, in fact, serves to illustrate how Hahnemann regarded many substances as medicinal and how there is no such thing as a wholesome medicine: they are all hurtful, producing primary and secondary effects. To demonstrate this point, in the 1803 essay "Der Kaffee in seinen Wirkungen" ("On the Effects of Coffee") Hahnemann wryly observes that, like tobacco, no one ever liked coffee the first time they drank it. He even describes a consummate coffee drinker upon waking in the morning: she has "the power of thinking and the activity of an oyster" ("On the Effects of Coffee," *Lesser Writings* 394). Its secondary effect is to leave "a disagreeable feeling of existence, a lower degree of vitality, a kind of paralysis of the animal, natural and vital functions" (394). Indeed, although Hahnemann does not include coffee in his remedy inventories, *coffea cruda* circulates today as a homeopathic remedy to counteract insomnia based on this after-effect of "lassitude and sleepiness."

Hahnemann was not alone in noting the two-part action of drugs, as the case of opium epitomizes. Although historians of homeopathy focus on Peruvian bark (because through self-testing it Hahnemann arrived at his resolution *similia similibus curentur*),[4] the example of opium in many ways better situates him within his era. Late-eighteenth-century physicians generally regarded opium as suspicious because the withdrawal symptoms of illness could be made to disappear by simply taking more of it: the concept of addiction was foreign to them.[5] Hufeland, for instance, coined the term "opium addiction" only in 1829. Instead, there was considerable debate about whether opium was a sedative or a stimulant, and which reaction preceded or was derived from the other. Already in 1679, Johann Jakob Waldschmied (1644–87) had indicated that opium could be a double-edged sword.[6] And in 1707 Georg Ernst Stahl (1659–1734), predating Hahnemann's criticism of the transitory, palliative effect of drugs, fought against the use of opium because of its merely transient lessening of symptoms. In his therapeutic nihilism, Stahl argued that it was better to let nature take its course. Hufeland, too, in coining the term opium addiction, had been opposed to its liberal prescription by the German followers of Scottish physician John Brown. Brown had recommended it as a prime stimulant to counteract conditions of debility or, in his terms, asthenic diseases.[7] As Andreas-Holger Maehle has pointed out in his chapter "Opium: Explorations of an Ambiguous Drug," by 1789 Georg Christoph Siebold (1767–98), and by 1793 Samuel Crumpe (1766–96), assumed that, "rather than different dosage alone, different *stages* of the action of opium were . . . responsible for the observed differences in effects" (164). Crumpe, writing in terms of Brown's system, argued that the "first stage was that of stimulation and excitement, which, as the body's excitability became exhausted, turned into the second stage" (165). Alexander von Humboldt also considered overstimulation to be the cause of the narcotic, sedating action of opium (166). Another scientist of the time, Johann Christian Reil (1759–1813), rejected opium to subdue a mentally ill patient because it created a fool of a different sort (*Rhapsodieen* 47). Not insignificantly, Brown, Crumpe, and Siebold, as well as numerous other physicians, including William Cullen (1710–90) (whom Hahnemann translated) and Friedrich Wilhelm Sertürner (1783–1841) (who in 1805 isolated the first opium alkaloid, morphine, giving birth to the science of pharmacology), conducted their opium experiments on themselves.

These views on opium share similarities with many of Hahnemann's principles and ways of knowing: 1. The debate centred on the primary and secondary effects of a drug and how one effect could instigate its opposite; 2. Even though opium was not yet seen as habit-forming, its transitory effects indicate the danger of a palliative drug; 3. Self-testing was regarded as the most reliable form of experimentation and, as Maehle also concludes, these "experiments were rather 'embedded' in speculation than used as crucial tests for hypotheses" (197). Furthermore, "there was not always a straightforward way from theory to therapy. Extensive therapeutic experiences with opium counted more to eighteenth-century medical practitioners than the latest interpretations and explanations of its actions" (197). Maehle's overview of physicians' experiences with opium help place Hahnemann's own self-experimentation with drugs, generally seen to be innovative, within a broader historical context. They also shed light on his understanding of objectivity and empiricism, as will be discussed in the next chapter.

What, though, did Hahnemann himself write about opium? He harshly criticized opium as the allopathic medicine of choice for coughs, diarrhea, vomiting, sleeplessness, cramp, and nervous conditions, conditions which would sooner heal themselves naturally than via opium – and which, if chronic in nature, would only become worse with its use. Opium thus serves as a primary example of the improper use of palliatives. Hahnemann wrote in the *Organon*: "The greater the quantity of the opium administered to suspend the pain, in the same degree does the pain increase beyond its primitive intensity when the opium has ceased to act." He then saliently illustrates the secondary effect of opium: "As in a dungeon where the prisoner scarce distinguishes the objects that are immediately before him, the flame of alcohol spreads around a conciliatory light; but when the flame is extinguished, the obscurity is then greater in the same proportion as the flame was brilliant, and now the darkness that envelops him is still more impenetrable, and he has greater difficulty than before in distinguishing the objects around him" (134–5, §69).

In the publication of the *Reine Arzneimittellehre* of 1830, Hahnemann had acknowledged the recent pharmacological discoveries of morphine and narcotine (codeine was only isolated later in 1832, thebaine in 1833, and papaverine in 1848). At the same time, though, he stipulated that these findings have no bearing on homeopathy: homeopathy needs to operate with complete, indivisible substances in their natural state (*Gesamte Arzneimittellehre* 1418). He also recognized, understandably given the circulating disputes, that it was harder to determine the effects of opium than those of any other drug (ibid. 1418). But this difficulty did not prevent him from stressing that opium is alone among substances for *not* producing any pain in its primary effects – thus making it unusable for the homeopathic treatment of pain (ibid. 1420). Instead, if at all, it is indicated for the opposite condition – for the loss of feeling. Thus, following the logic of *similia similibus curentur*, in the *Organon* he recommends a high potency of opium in chronic conditions where the patient, having suffered for so long, is in a "depressed state of . . . sensibility." Opium "will remove the torpor of the nervous system, and then the symptoms of the disease develop themselves plainly in the re-action of the organism" (182, §183).

In short, as the example of opium illustrates, the concept of the biphasic action of drugs was not unique to Hahnemann. But Hahnemann did foreground it as a general rule about medicinal substances in 1796, and it provided the key to *similia similibus curentur*. Hahnemann then explained the significance of primary and after-effects more extensively in the 1805 essay "Heilkunde der Erfahrung" ("The Medicine of Experience"):[8] "In order therefore to be able to cure, we shall only require to oppose to the existing abnormal irritation of the disease an appropriate medicine, that is to say, another morbific power whose effect is very similar to that the disease displays," and "it is only by this property of producing in the healthy body a series of specific morbid symptoms, that medicines can cure diseases, that is to say, remove and extinguish the morbid irritation by a suitable counter-irritation" (451). Unlike drugs working according to the principle of *contraria contrariis*, which can aggravate and intensify the original disease, the cure according to *similia similibus* produces a slight aggravation only *resembling* the original disease. This slight aggravation causes the body's own vital force to overcome the original illness, resulting in a permanent cure. Here it is important to recognize the distinction between a *symptomatic* (i.e., palliative) *treatment* and the homeopathic *treatment of symptoms* intended to fully cure a disease. By 1811 in the first edition of the *Reine Arzneimittellehre* Hahnemann referred to the secondary symptoms elicited by a drug as their healing effects. Another way of putting it is to say that the secondary effect leads to the fast, permanent restoration of vigour.

In the *Organon* the term "secondary effect" thus comes metonymically to stand for the reaction and reassertion of the vital life force in the living organism: "Our vital powers tend always to oppose their energy to this influence or impression [of the medicine or primary effect]. The effect that results from this, and which belongs to our conservative vital powers and their automatic force, bears the name of *secondary effect*, or *re-action*" (130, §63).[9] For Hahnemann, the homeopathic remedy exercises a "more potent power" that allows the body to regain equilibrium and overcome the initial corporeal affliction. Hahnemann also visualized the homeopathic healing process in terms of a facsimile, copy, or artificial replica of the original disease that the remedy creates; this facsimile then substitutes for the original and disables it. In the *Organon* Hahnemann rephrased the curative action: "A remedy . . . closely resembling the natural one against which it is employed . . . excites . . . the artificial disease . . . [and], by reason of its similitude and greater intensity, now substitutes itself for the natural disease" (171, §148; see also "The Medicine of Experience," *Lesser Writings* 455). Key to homeopathic theory is the belief that two competing stimulants cannot reside in the body at the same time without one overcoming and cancelling out the other (ibid. 447).[10]

Although, of course, Hahnemann did not use this comparison (and I shall consider in a moment the ones he does deploy), one way of illustrating how *similia similibus curentur* leads to recovered equilibrium is to consider the martial arts: they too are based on a theory of *interface* with an opponent, not *confrontation*. One uses the resistance of one's own body to restore balance, thereby converting enmity into a new symmetry. In homeopathy, too, what initially appears to be enmity in the guise of a pharmakon (enmity because the drug creates symptoms similar to the disease) actually prompts the body to regain equilibrium. As in the martial arts, the foe is converted into the friend, the enemy into the ally. In the martial arts all depends on subtle, clever positioning – in homeopathy on the effective, imperceptible dose. Homeopathy thus does not operate, as in heroic medicine, on metaphors of overcoming blockages, obstacles, and resistance, all of which imply stasis and a retrograde confrontation. Instead, it implies motion, impulsion, and pliancy in the restoration of health.

Johann Ernst Stapf (1788–1860), Hahnemann's disciple and editor of the first homeopathic journal, prefaced each issue with the famous lines from act 1, scene 2 of Shakespeare's *Romeo and Juliet*: "one fire burns out another's burning, / One pain is lessen'd by another's anguish; / Turn giddy, and be holp by backward turning; / One desperate grief cures with another's languish: / Take though some new infection to thy eye, / And the rank poison of the old will die." It is not surprising that, to illustrate the principle of *similia similibus curentur*, Hahnemann too would turn to similes, though ones of his own devising.[11] They were no less vivid than Shakespeare's:

> Why does the brilliant planet Jupiter disappear in the twilight from the eyes of him who gazes at it? Because a similar but more potent power, the light of breaking day, then acts upon these organs. With what are we in the habit of flattering the olfactory nerves when offended by disagreeable odours? With snuff, which affects the nose in a similar manner, but more powerfully . . . In the same manner, mourning and sadness are extinguished in the soul when the news reach us (even though they were false) of a still greater misfortune occurring to another. (*Organon* 106, §26)

Hahnemann conversely illustrates the paltry effects of the palliative: "The tears of the mourner may cease for a moment when there is a merry spectacle before his eyes, but soon the mirth is forgotten, and the tears begin to flow again more freely than ever" (*Organon* 134, §69). In addition, he noted that *contraria contrariis* led to the misfortune expressed in the phrase "Spare the rod and spoil the child." Attempting to mollify a bad child would only make him worse, for instance, when parents "imagine that a sweet cake is the remedy for [a child's] peevishness and rudeness . . . The poor parents have now recourse to other palliatives: toys, new clothes, flattering words – until at length these are no longer of any avail." Had they had used the rod only once, although for the first half hour the child might become more "unruly, bawl and cry somewhat louder, . . . it would subsequently become all the more quiet and docile" ("The Medicine of Experience," *Lesser Writings* 456).

One of Goethe's early plays – predating, with its third version of 1790 Hahnemann's essay by six years – illustrates well the application of *similia similibus curentur* in this emotional, disciplinary, yet also curative realm. At the beginning of *Lila*, the title character's sisters lament the horse medicine their sibling is subjected to in an effort to cure her melancholy. They are mistrustful of physicians who only confirm in the patient what they want to see: they operate and apply enemas or electrocution, only to see what they had predicted – the patient's grimaces (WA 1.12: 46). A new doctor, though, announces a new, gentler approach: Veragio decides to cure Lila's fantasies by using fantasy. The household sets about to engage Lila in her imaginative self-enclosed world by making her become active in rescuing her husband and sisters, who pretend they are under the magic spell of an ogre. Lila, gathering courage, says: "Perhaps I am destined to free you and make you happy. The heavens often unite the unhappy to relieve their dual miseries" (WA 1.12: 71). She decides to deliver herself to the chains her family are also subject to and thereby save them. Summarizing this theory of like curing like, one character by the name of Friedrich sings at the close: "What was taken away by love and imagination will be brought back by love and imagination" (WA 1.12: 85). Indeed, in the following years, the Halle physician Johann Christian Reil, who instituted the study of mental illness in his work *Rhapsodieen über die Anwendung der psychischen Curmethode auf Geisteszerrüttungen*, advocated a theatre cure: sickness of the nerves that have their origin in fantasy need to be cured by fantasy. Seeing one's problems embodied on the stage leads to their relief and to healing.[12]

Hahnemann's Semiotics: Aligning Similarities

In order to pursue further the first pillar of homeopathy, the Law of Similars – which by 1807/8 Hahnemann was calling a law of nature and the only road to healing (Schmidt, "Samuel Hahnemann und das Ähnlichkeitsprinzip" 151) – it is necessary to interrogate how he proceeded in developing and ratifying this "law" after the inaugural essay of 1796 (where it was only mentioned as a heuristic "principle"). What was Hahnemann's concept of symptoms that allowed him to select and compare them? How did he define disease, and what common names for diseases would he use, if any? What method did he use to come up with his remedies? How did homeopathy compare to other eighteenth-century diagnoses of the ill body?

To answer these questions, I wish to identify two separate moments in Hahnemann's thought – the difference between his semiotics (how he observed, recorded, collected, and compared signs syllogistically) and his conjecture in selecting a remedy (how he isolates a particular sign to solve the case). The former is based on a pure process of cataloguing and cross-referencing: it is based on how signs refer to other signs.[13] The latter is based on how he singles out a noteworthy symptom to lend peculiar weight to it and how this symptom clinches his decision about which remedy to select. The opposition here is between rational systematization and inspired reading.

Throughout the course of his career, Hahnemann set about studying the reactions that substances produced in a hale and hearty person, reasoning that, when this reaction mimicked a true disease, the homeopathic remedy was found. The task he undertook over the course of his life was to determine via close observation of healthy individuals – most often himself but also his family and students – what symptoms drugs produced. Perversely, then, homeopathy is based on provoking the body to illness and punctiliously recording the results rather than documenting the pathway to health. Hahnemann first recorded these reactions in protocol books.[14] He subsequently catalogued them in repertories (by symptom) and *materia medica* (by remedy) – a written mediatization of the body resulting in vast compendia. His procedure was one of closely observing indicators – first in the healthy and then in his patients – and matching them so as to come up with the appropriate homeopathic remedy. His idea was that one could not truly know what occurred in the human body, but that it presented external signs to be read: "The internal essential nature of every malady, of every individual case of disease, so far as it is necessary for us to know it, for the purpose of curing it, expresses itself via the symptoms" ("The Medicine of Experience," *Lesser Writings* 443). For him, it was not that a drug would overpower a disease, but that one symptom would overcome the other. As he writes in the *Organon*:

> The particular medicine whose action upon persons in health produces the greatest number of symptoms resembling those of the disease which it is intended to cure, possesses, also, in reality (when administered in convenient doses), the power of suppressing, in a radical, prompt, and permanent manner, the totality of these morbid symptoms – that is to say – the whole of the existing disease. (105, §25)

To restate the issue, it was not that at the root of symptoms was a disease that needed to be fought off via drugs: homeopathy operated instead syllogistically by finding similarities between sets of symptoms and concluding that the substance that produce the first set could cure the second. The underlying principle of homeopathy was thus a semiotic one, based on an association and compilation of signs. Hahnemann's process was first to distinguish indications of illness in recording what he observed and subsequently to constitute them as signs by comparing them to other signs. Unlike the early nosologists, such as Boissier de Sauvages (1706–67), Carl Linnaeus (1707–78), and Cullen, who began to *classify* diseases into families and tables, Hahnemann *compiles* symptoms and remedies.

Hahnemann created this coverage in the compendia of the *Fragmenta de viribus medicamentorum, Reine Arzneimittellehre,* and *Die chronischen*

Krankheiten. Christian Lucae and Matthias Wischner list 125 remedies in sum in their tabular overview (*Gesammelte Arzneimittellehre* 2018–24), remedies that form the substantial basis of the homeopathic repertoire today. One cannot stress enough how immense and bewildering this textual reproduction of bodily illnesses are. Lucae and Wischner, for instance, count over 9000 printed pages. The 1796 founding essay alone mentions 65 pharmaka. The *Fragmenta* was a motley grouping of symptoms under the name of one drug. Helen Varady notes that it received little notice, perhaps not only because the Latin was idiosyncratic, but also because it was hard to decipher: although the first volume refers to only 27 remedies, it listed 2647 symptoms derived from self-testing, plus 1698 more. Although volume 2 had a register of symptoms, it was still difficult to navigate. Hufeland additionally criticized it for the lack of information on the testers, on the effects of certain important medications, and on the dosages prescribed ("Fragmenta de viribus" 224–5). With its sixty-three remedies, the six-volume first edition of the *Reine Arzneimittellehre* was not much different in its heterogeneous assemblage, except that the symptoms for any particular remedy were listed starting with the head and progressing to the toe. *Die chronischen Krankheiten* altered the system of categorization only insofar as emotional symptoms were listed first. Yet another collection, though unpublished, was Hahnemann's own large repertories, where symptoms were cut and pasted, in the minutest handwriting, in alphabetical order (see figure 1).[15] Here the various ways in which headache alone can manifest itself take up pages 272 to 306 (see figure 2).

Yet despite the incongruities and incommensurables between symptoms which resulted from this voluminous amassing, Hahnemann saw himself throughout his career as setting up concordances that spanned across discordant particulars. He purported to couple data, establish equivalences, and determine reciprocities. Homeopathy was designed as an integrated network of resemblances and coordinates that arose across separation. The Law of Similars governed how it operated. It conveyed an image of truthfulness and empirical accuracy.

On what symptoms of illness, precisely, would Hahnemann concentrate? The general symptoms of importance that he would note in his patients would be: 1. Their mental state, attitudes of will, determination, and aversions; 2. The effect of different temperatures and times of day (periodicity) on their symptoms; 3. The sides (left or right) of the body affected; 4. Their cravings for or aversions to particular food substances, such as sweets, fats, and salt, as well as their thirstiness

Figure 1 Hahnemann's unpublished repertory of symptoms. Courtesy of the Institute for the History of Medicine of the Robert Bosch Foundation, Stuttgart.

or lack thereof; and 5. The alternation of complaints. This last point is significant insofar as Hahnemann believed that the correctly chosen remedy would cure a host of ailments; thus, in §42 of the first edition of the *Organon*, he noted how the whole body reacts in sympathy to an injured organ. In terms of localized symptoms, he looked at such indicators as the tongue, pupils, pulse, skin temperature, and sweat outbreaks.[16] He would take into account family circumstances, tragedies, sorrows, and financial worries. He had no compunction about inquiring about sexual intercourse and masturbation. With women, he noted their monthly periods. Auscultation and percussion, however, were not yet common practice.[17] As the above makes clear, Hahnemann did not physically examine a patient but relied on his or her verbal testimony (Fischbach-Sabel 76).

In her study on the *Fragmenta de viribus medicamentorum*, Marion Wettemann notes that, although only at a later date did Hahnemann say

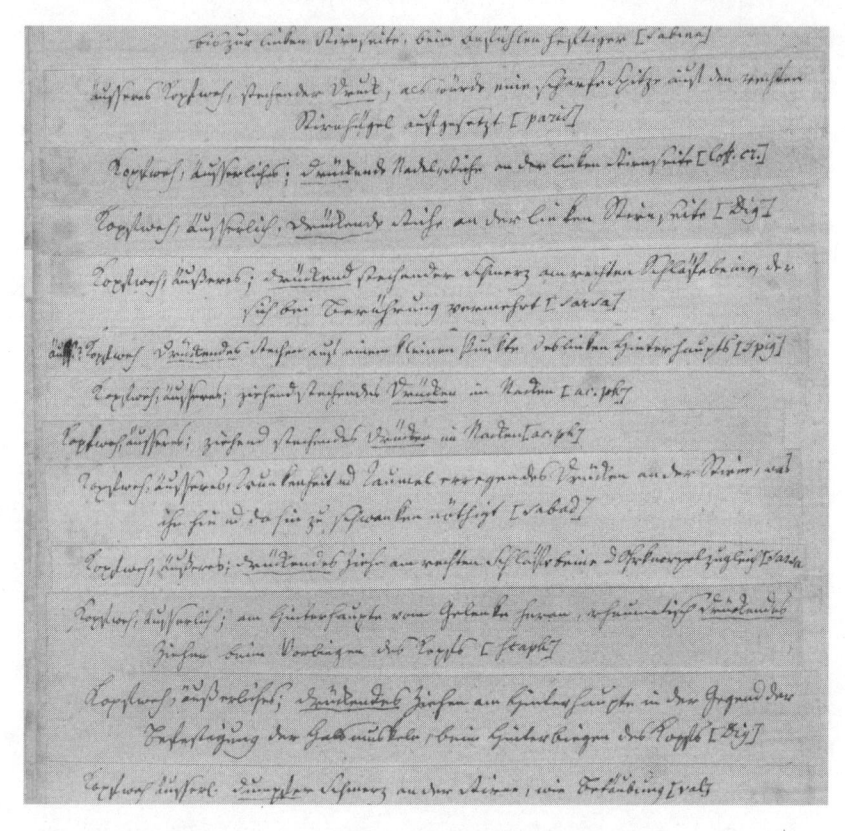

Figure 2 An excerpt from the entry on headache in Hahnemann's repertory. Courtesy of the Institute for the History of Medicine of the Robert Bosch Foundation, Stuttgart.

what characteristic symptoms a substance produced, already here he singles out a few such symptoms, for instance, the lack of thirst when *pulsatilla* is indicated (34–5). In her investigation of the case journals that arose around the time of the *Fragmenta*, Varady similarly notes specific remedies for emotions, although their causes might vary: for instance, he dispensed *chamomilla* for anger and *ignatia* for shock. Hahnemann would even substitute the word "fear" for the medication in his notes, prescribing "2 sugars of fear" (39–40). The attention to symptoms involved subtle interpretation. For instance, because medications could distort symptoms, Hahnemann often felt he needed to wait until their effect had worn off, during which time he would prescribe placebos. In his later work on chronic diseases of the skin he noted that outer symptoms could even be suppressed and remain unseen (see *Organon* §§202 and 203, 5th ed.).

In terms of specific illnesses and diseases, Varady notes in reference to the case journals between 1803 and 1806 that colds played a minor role, with only 83 out of 1644 cases, comparable to the number of cases with urinary problems (209). Hahnemann also listed asthma, diabetes, pleurisy, migraines, rabies, hemorrhoids, rheumatism, malaria, and consumption, among other ailments.[18] But because Hahnemann believed that disease differed from person to person, these names are only used as a short form for convenience and do not exclude his examining a patient's entire ills. For the most part, as Ute Fischbach-Sabel observes in her commentary to the case journal of 1830, he does not even use the names of illnesses. And when he does, he does not let the name determine the choice of remedy (20). Indeed, morbidity would be reflected in the *totality* of symptoms, not in any key ones unique to a disease that, according to other early-nineteenth-century physicians, guided a diagnosis. As Haehl succinctly summarizes, "He declined to consider the name of a malady, based only on one prominent symptom, just as he declined to allow a medicine to be used for one or several symptoms simply with the intention of suppressing them (symptomatic method of curing)" (298).

How, though, did Hahnemann's conception of disease and way of reading symptoms compare to those of other physicians of his day? And how did they evaluate his semiotics? Before answering these basic questions directly, one needs to step back to consider, however briefly, the preceding 150 years of medical advances.[19] Thomas Sydenham (1624–89) could be called the father of nosography and nosology insofar as he wanted to know the species of illnesses, just as one could know the species of plants: as Karl Rothschuh puts it, he focused on *Krankheitsgeschichten* (the progress of diseases) instead of *Krankengeschichten* (the narratives of patients) (*Konzepte der Medizin* 166). Still, Sydenham is classified as an empiricist who trusted his senses. Robert Boyle (1627–91), by contrast, wanted to know what went on behind the empirical senses. This conflict between the empiricists and the rationalists continued for almost two hundred years. For Herman Boerhaave (1668–1738) and Friedrich Hoffmann (1660–1742), for instance, bedside observation was not enough: the physician needed to know the causes of disease and explain how the body functioned.[20] They were adherents of seventeenth- and early-eighteenth-century mechanical philosophy and believed one needed to avoid metaphysical speculation.[21] Hoffmann, in particular, saw the body in terms of hydraulics, statics, and hydrostatics. As Roger French put it, "The anatomy of fine structure shows that the body is a machine, with its columns, levers, springs, wedges, pulleys, bellows, sieves, and presses. The doctor who recognizes these and knows the laws of motion, concludes Hoffmann, will also recognize how the machine can break down, and so knows the essence of pathology" (98).

In contrast to Boerhaave and Hoffmann, in the Montpellier school of medicine in the eighteenth century, a discourse of vitalistic forces rather than mechanisms took hold. Its founder Sauvages was influenced by Stahl's theory of an *anima* regulating the body; yet at the same time Sauvages advocated precise empirical observation in order to collect and classify illness, as in his 1763 taxonomical *Nosologia Methodia*. Théophile de Bordeu (1722–66), discontented with the Stahlian metaphysical framework adopted by his predecessor, located sensibility in specific organs and physiological systems. Others, such Pierre Georges Cabanis (1757–1808), Xavier Bichat (1771–1802), and Philippe Pinel then furthered the recognition that the physician must detect the causes of disease. Theirs was a science of etiology which refined a methodical-empirical approach to understanding phenomena. It would supplement bedside practice; in fact, such knowledge often took precedence over a cure. Bichat investigated pathological anatomy and ascertained that diseases derived from tissue lesions. Pinel studied the brain. And once Bichat and Pinel looked *inside* the body to its tissues, the era of classical empirical, *external* observation of the patient ends.

In his resistance to explaining what transpires inside the body and in his exclusive focus on symptoms in order to determine the remedy that cures, Hahnemann is a child of this older eighteenth-century empiricism that was waning and inevitably could not compete with the rise of pathological anatomy in the nineteenth century that focused on specific diseases as clinical entities rather than on the sick individual.[22] Restated, as it progressed after 1800, medicine could not share Hahnemann's twofold belief that there was no such thing as localized illness and that symptoms governing the entire body needed to be recorded instead. As opposed to treating the *sick individual*, the nineteenth century became more adept in studying *individual diseases* that surfaced in the larger hospital clinic. Yet despite their eventual decline, the advocates of empiricism meanwhile counted among the luminaries of eighteenth-century medicine. Hahnemann's scepticism regarding the practical implications of pathology and knowledge of internal causes, for instance, reflects that of the Vienna School where he had studied.[23] Roy Porter singles out Albrecht von Haller (1708–77) as another example: "Like Newton when faced with the phenomenon of gravity, Haller believed that the causes of such vital forces were beyond knowing – if not completely unknowable, at least unknown. It was sufficient, in true Newtonian fashion, to study effects and the laws of those effects" ("Medical Science" 145). If it displayed no therapeutic results, there seemed to be little purpose in merely describing, naming, and classifying a disease or in knowing its first causes. Again, Roy Porter writes on eighteenth-century medicine: "Questions as to the true causation (*vera causa*) remained highly controversial. Many kinds of sickness were still attributed to personal factors – poor stock or physical endowment, neglect of hygiene, overindulgence, and bad lifestyle. This . . . made excellent sense of the uneven and unpredictable scatter of sickness: with infections and fevers, some individuals were afflicted, some were not, even within a single household. It also drew attention to personal moral responsibility and pointed to strategies of disease containment through self-help" ("Medical Science" 150). Hufeland and Hahnemann both were physicians of the time who highlighted the prerequisite of a healthy diet and lifestyle.

All in all, then, because illness was unpredictable and uneven in its appearance from person to person, Hahnemann felt he had to dispense with both *physiology* (anatomy of the body in a healthy condition) and *pathology* (the origin, nature, causes, and development of diseases) in order to focus instead on the *semiology* of illness which would attend to the specifics of each individual as well as on *dietetics*. This semiology involved the *anamnestic* recording from patients to the exclusion of a *diagnostic* conclusion about a disease and of its *prognostic* course.[24] Thus, Hahnemann is able to summarize that via careful, thorough, external sensory observation "the physician will succeed in depicting the pure picture of the disease, he will have before him *the disease itself*, as it is revealed by signs" ("The Medicine of Experience," *Lesser Writings* 447).[25] Today, however, we would say that only a para-science studies effects not causes.

Three further references suffice as illustrations of the prevalent eighteenth-century conviction that a physician should not speculate about the inner workings of the body and instead concentrate on experience and observation. All three stem from 1795, a year before the inaugural date of Hahnemann's first essay on *similia similibus*, thus offering a synchronic snapshot of the tradition into which he fits. In the preface to the first volume of his *Journal der practischen Arzneykunde und Wundarzneykunst* (where Hahnemann published his "Versuch" a year later), Hufeland stipulated that "each observation, accurate and true to nature, each contribution to improve the knowledge of healing treatments, each practical note, however short, shall be gratefully received" (preface to the *Journal* iv–v). The second example was penned by Johann Christian Reil in his essay "On Vitality": "I shall not comment on the nature of this matter, whether it is warmth, electricity, oxygen, etc., substances with known effects on the body; nor shall I comment on the relationship between such substances and the rough matter, of the changes they cause in the animal body, because our experience is too limited" (15). As will be seen more thoroughly in the third chapter, to posit a vital life force, as do Reil, Hufeland, and Hahnemann, means to dispense with eighteenth-century iatrochemical and iatromechanical explanations of the body, that is to say certain schools of medicine that maintained physiology could be elucidated by chemistry or physics. The third example comes from the opening of *Allgemeine Bemerkungen über die Gifte und ihre Wirkungen im menschlichen Körper* (*General Observations on Poisons and Their Effects on the Human Body*) where its author, Carl Christian Heinrich Marc (1771–1840) quotes Haller: "The inner being of nature no created mind shall comprehend, / happy he to whom the outer shell is revealed" (n.p.).

Despite these similarities to Haller, Störck, Hufeland, Reil, and Marc, though, as indicated at the beginning of this chapter Hahnemann clearly was at odds with the predominant medical treatments of his day. As should now be apparent, the reason for his polemic was that eighteenth-century physiological discoveries did not immediately see the fruits of therapeutic innovation and success; in addition, many of the old Galenist therapies intended to govern the flows in the body were still in use.[26] Of course, neither the presence of bacteria nor their cure in penicillin had yet been discovered. Filling this vacuum, then, around 1800 was the speculative medicine of Brown (to whom I shall return in chapter 2), on the one hand, and that of Hahnemann, on the other. However they diverged, both physicians claimed to address therapeutic dead ends by claiming to prescribe medication and dosage accurately.

What, though, of the broader picture apart from medicine that allows us to see Hahnemann's semiotics as typifying his age? One can compare his semiotic procedure to what Michel Foucault categorized as typical of the scientific method in the Classical Age. Foucault argues that, for instance, natural history as it arises in the seventeenth and eighteenth centuries is "established within the apparent simplicity of a *description of the visible*" (*Order of Things* 137). The object is constituted or "provided by surfaces and lines, not by functions or invisible tissues. The plant and the animal are seen not so much in their organic unity as by . . . visible patterning" (ibid). To be sure, Hahnemann did regard the human body as an organic unity whose equilibrium, once disturbed, would result in malady. But it was not the *internal* circulatory, muscular-skeletal, digestive, or nervous systems that, according to homeopathy, could break down under disease so that the physician could analyse its etiology and development and then treat it; rather the *external*, or what Foucault calls "visible patterning" in the guise of symptoms were the solution to the cure.

Key here is that the somatic and psychic manifestations of disequilibrium do *not* reference the body – conceived as a unity of organs, tissues, and organic systems – and hence do not reference a disease that can be named. The similarity between effects indeed negated the need to find a cause for disease. What mattered was the accumulation and compilation of observable symptoms, not a semantic chain that would indicate an etiology or chart a pathology. As a new science, in its self-generating productivity, homeopathy creates its own encyclopedia of symptoms and its own pharmacopeia that its practitioner consults. Comparable to the study of botany, the rule of similars allows for abundance and difference. Homeopathy, much like natural history, thus "traverses an area of visible, simultaneous, concomitant variables, without any internal relation of subordination or organization" (*Order of Things* 137). Hahnemann would therefore see the role of his *Reine Arzneimittellehre*, inasmuch as it is a compilation, "as [a] contribution to the collective store of observations" (Pickstone 68). In addition, by neatly exposing pairs of symptoms, Hahnemann epitomizes what David Wellbery sees as the Enlightenment desideratum of a "fully perfected sign system" (42). Insofar as Hahnemann thereby claims to have found nature's remedy to illness, in Wellbery's words, "*nature is recovered in the form of a completely transparent language that is equivalent to divine cognition*" (42).

Hahnemann's method was also comparable to that of eighteenth-century philosophers, for whom "such a method [hierarchies based on maximal numbers of 'characteristics' – all treated as equally important] approximated the mental processes of 'association' which were fundamental to learning" (Pickstone 70). Rothschuh points out that reasoning by analogy was important, too, for the medical empiricists – the comparison of observations, the collection of data leading to classification, and the effort to find a correlation between special diseases and the specifics to heal them (*Konzepte* 174). Outside the realm of medical semiotics, but still sharing the belief that corporeal signs naturally and transparently revealed themselves, is the work of the famous eighteenth-century German physiognomist Johann Christian Lavater (1704–1801). Like Hahnemann, he too idiosyncratically but systematically compiled volumes of observable, exterior signs on the body, though he focused instead on how the interior, moral intelligence of a person was written on his or her face. As with Hahnemann, every external corporeal detail was significant: "Consequently, there is not a wrinkle, not a tiny wart, not a hair on the human body that is not already, in a physiognomic sense, an unmistakable, yet not unraveled, sign for the open-minded" (*Aussichten in die Ewigkeit* 3:54).

But how were Hahnemann's semiotics and nomenclature received by his contemporaries? Three major contemporary reviews of the *Organon* are indicative of the change medical interpretation of pathological signs underwent by the start of the nineteenth century. In his two-volume review of Hahnemann's first edition of the *Organon* in 1810, August Friedrich Hecker (1763–1811) observed with regard to Hahnemann's symptomology: "The peculiar and characteristic micrology and tedious scrupulosity can also be found in the rather strange description of the symptoms" (37). "All that is presented side by side, upside down and inside out, without ever revealing how, where and when the remedies do this, that or another" (45). Hecker's critique suggests that by the beginning of the nineteenth century medical experts are no longer content to see the sheer listing of symptoms, combined with a disregard for how medicines work. In addition, Hecker took Hahnemann to task for not deriving the symptoms from a firm medical cause (221). Hahnemann's strict eighteenth-century concentration on observation thus leads to "a genuinely ridiculous empiricism" (244). Johann Christian August Heinroth (1773–1843), who composed an *Anti-Organon* in 1825, notes that physicians today "observe the symptoms to identify the disease, i.e. to identify the pathological or abnormal mutations of the conditions inside the body" (26); not to do so is to remain stuck half-way. Of Hahnemann's empiricism he says that he merely wants to think with his senses (27). Instead Heinroth favours a semiotics as "the crown of existing medicine" that leads to diagnostic, even morphological insights (85). Hufeland is more generous in his 1826 essay, republished as *Die Homöopathie* in 1831. He mentions several advantages of homeopathy, including that physicians in the future will pay more attention to the semiology and symptomology of disease. But predictably for the era, and similarly to Heinroth and Hecker, he counts among the disadvantages that Hahnemann does not look at the *causa morbi proxima* and that, like Brunonianism, homeopathy retreats from scientific advances in anatomy, physics, chemistry, pathology, and etiology. Current medical diagnosis, he says, singles out the symptoms that are important, non-incidental, and common to all patients with a disease.[27]

Finally, Hahnemann's semiotics also invite contrast – if for no other reason than he also used the phrase *similia similibus curantur* – with that of a much earlier medical thinker, Theophrastus von Hohenheim, otherwise known as Paracelsus (1493–1541). In referring to Paracelsian thought, Foucault notes that "the world of similarity can only be a world of signs" (*Order of Things* 26); this world view would also be true of Hahnemann. But whereas Paracelsus ascribed to the signature of things, according to which "even though he has hidden certain things, [God] has allowed nothing to remain without exterior and visible signs in the form of special marks" (qtd. ibid. 26),[28] Hahnemann does not refer to a magical analogy that reveals the workings of God. His system of analogy is without reference to the sympathy between microcosm and macrocosm. Hahnemann did not read the medicinal purpose of a plant by virtue of how it looked. In other terms, it is no longer the plant itself via its appearance that suggested an affinity with the cure it could bring about; it was solely the effect of the plant on the human body that Hahnemann examined and recorded. Thus one does not find in Hahnemann, as in Paracelsus, a belief in a vast system of signatures that revealed the invisible workings of a divinely inspired and created universe. He does not believe in an analogical *universe*, only an analogical *system* for registering symptoms. And certainly planetary movement was not aligned, as it was in Theophrastus von Hohenheim, with processes of healing.[29] Hahnemann believed that he was documenting observable empirical positivities. To repeat, he created a closed system, typical of eighteenth-century thought, whereby signs referred to other signs, not to the macrocosm and its divine order. Rather than the Paracelsian vertical signification, in homeopathy signs refer horizontally to each other. Above all, what mattered foremost for Hahnemann was the *collection* of signs and remedies; the only *patterning* that would be significant was how the primary effects of a drug would match those of an illness.

That said, homeopathy does resemble the older tradition of alchemy for which Paracelsus stands in its belief that the elemental, essential action of a substance could be extracted and transmitted. According to alchemy and homeopathy, the latent, spectral energy of an ingredient unfolds and is clarified through a process of distillation. In alchemy, the spirit of metals was supposed to have been released from its base surroundings upon heating: through the homeopathic dilution, the spiritual force in the tincture was likewise claimed to be refined and released. Finally, although homeopathy is not about finding the elixir of life or extracting gold from base metals, the homeopathic operation similarly involves only a few particles of matter.

Hahnemann's Inspiration: Isolating Dissimilarity

In chapter 3 I shall have more to say about the transmutation of substance in the infinitesimal dose. But here, with regards to semiotic interpretation, the topic of magic sneaks in via another, less obvious door. In his essay "Lehre vom Ähnlichen" ("Doctrine of the Similar") Walter Benjamin speaks of the "moment of birth" in the perception of similitude: correspondences appear to one suddenly, arising at the ingenious spark of inspiration (2.1: 206). He compares this occurrence to the flash of insight that comes to the astrologist in seeing the conjunction of two stars, when he perceives a third term or special meaning in their constellation. This magical, unanticipated instant leads Benjamin to work out a concept of a supersensual similarity ("Begriff einer unsinnlichen Ähnlichkeit"; 2.1: 207).[30] In other words, in counterpoint to the establishing of similarities stands the pivotal but paradoxical idea that what actually grounds comparison is something unexpected and not grounded in sense perception. Benjamin offers the examples of onomatopoeia and graphology as beliefs in an innate but non-material correspondence between a sign and that to which it refers.[31]

This concept of a third, sudden, dissimilar element unexpectedly facilitating analogy is one that invites investigation with respect to *similia similibus curentur*. As concluded in the previous section, the semiotic system or tabling of symptoms that Hahnemann set up endeavoured to be thoroughly systematic, based on close observation and the recording of data, as well as a contribution to the compiling of facts. Walter Benjamin's insight into the operations of similitude, though, suggests a very different mode of operating and encourages us to search for a third, inspirational, and unexpected element undergirding Hahnemann's system of analogies. Indeed, as we shall see in this section, the moment that clinches the decision for Hahnemann in determining which remedy to offer a patient is both what enables the comparison and, at the same time, threatens to disrupt the symmetrical order.

To state it differently, two divergent tendencies exist as ways of making meaning in Hahnemann's medical system. The one strives to list and catalogue symptoms based on their similarity; it is indebted to an eighteenth-century belief in taxonomical organization. The second tendency, running counter to the first, is a principle of the absurd, chaotic, and exceptional, in other words, Benjamin's exceptional, even nonsensical moment that grounds the comparison. This second tendency resembles less eighteenth-century collecting than Romantic inspired reading. It can be traced in Hahnemann's concepts of the unusual symptom and disease as unique to each patient. The first requires a semiotic alignment of signs based on evident parallels, while the second depends on the ingenuity of the individual reader to single out the pertinent, telling sign, in other words, a kind of divination. That both "ways of knowing" – to use the phrase coined by John Pickstone – existed simultaneously is not surprising. Samuel Hahnemann straddles the eighteenth and nineteenth centuries. Extremely learned and fluent in several languages, he embodies the eighteenth-century savant. And yet he also reflects beliefs in organicism and vitalism that we have come to associate with Romanticism. In sum, he borrowed from contradictory paradigms to construct his own salient and unique philosophy of medical treatment.

For each remedy in the *Reine Arzneimittellehre* Hahnemann lists a vast number of symptoms of the body, mind, and disposition which it can treat and often the time of day in which these symptoms appear. As noted in the preceding section, the portrait for each remedy is based on an accumulation of symptoms, which is to say that Hahnemann neglects either to exclude or to prioritize them. His method is synchronic in the sense that the duration, succession, frequency, and cessation of symptoms – their diachronic aspect – is not noted. For instance, more than forty pages are devoted to the remedy *nux vomica* and its various indicators such as vertigo, headache, smarting in the eyes, swelling of the gums, ringing in the ears, toothache, looseness of the teeth, heartburn, nausea, pricking pain in the hepatic region, flatulence, burning or itching while urinating, erection of the penis after the midday sleep, nocturnal cough, bloody nasal mucus, asthma, sudden powerless of the arms, frightful visions in dreams, and yawning accompanied by weepy eyes. Quite understandably if suffering from all these symptoms, the *nux vomica* patient also exhibits extraordinary anxiety, crossness, sadness, reproach of others, even mistakes in speaking and writing. This exhaustive coverage as well as listing, in which no detail is omitted of Hahnemann's investigations into *nux vomica* over several years, gives the impression of an intentional lack of hierarchy of symptoms as well as an asystematic presentation. Conceivably, there is no limit to the potential listing of symptoms, because, as we have seen, it is not the goal of the physician to arrive at a diagnosis, pathology, or nosology. In other words, there never arrives the instant at which Hahnemann, in compiling the provings, determines that enough symptoms have been recorded.

The only thread that joins the symptoms is the listing (at least starting in the *Reine Arzneimittellehre*) from head to toe, not any interpretation of the symptoms or their relation to one another. Put succinctly, this accumulation of several disjointed moments of a disequilibriated body threatens to collapse the Law of Similars. This law attempted to create order by drawing parallels between signs in two separate human bodies, rather than between warning signs in one body. The single body thus houses chaotic, isolated, nonstratified symptoms that in fact tyrannize it as incomprehensible illness. Whether one consults the *Krankenjournale* (to repeat, where the patient symptoms are recorded) or the *Reine Arzneimittellehre* (where the symptoms produced on the healthy person are recorded), each individual body presents a bewildering, cacophonous encyclopedia of symptoms. The bodily and psychic indicators of illness that are catalogued and recommended for each remedy seem infinite and unrelated, as if we were truly speaking here of a Deleuzian "body without organs," that is to say, a body without any unifying systems, be they digestive, nervous, circulatory, and so on. Such a body not only expresses its uniqueness through a plethora and mingling of affects or what Deleuze and Guattari in *A Thousand Plateaus* call "intensities,"[32] it is purportedly also, as we shall investigate in more detail in chapter 3, exquisitely tuned to respond to the minuscule homeopathic dosage. Operative in Hahnemann's system, in his note taking, and in his own patient's letters is thus less a regulatory, disciplinary monitoring of the body than a dissolving of self in the proliferation of discrete symptoms and the resonances of the remedies. The coordinates are incalculable, and the results are unverifiable. Important instead are the responses of the physiological and psychological bodies, each person infinitely diverse from the next.

But if the homeopath merely jots down the symptoms that a patient relates to him and is even encouraged to refrain from interpreting them, how then is the cure to the maladjustment in the body to be found? What results in the selection of a cure, I would like to argue, is comparable to Benjamin's concept of the suddenly intuited similarity. What enables the decision about what remedy to select is the bizarre, unanticipated moment. The lynchpin in deciding upon a treatment was based on Hahnemann's notion that each individual patient was unique, hence, that, despite similarities with other patients, what singled out for the physician the choice of a cure is what made the patient stand out from all other cases. In short, paradoxically only the dissimilar could enable the workings of the Law of Similars.

Hahnemann criticized allopathic medicine for attempting to reduce all individual cases to one disease, whereas he saw each individual case as unique. Diseases are infinite in number, he wrote, "as diverse as the clouds in the sky" ("On the Value of the Speculative Systems of Medicine," *Lesser Writings* 504). In striking contrast to medicine as practised later in the nineteenth century, he insisted that it was always the person with the disease who was treated, not the disease itself. "Each case of the disease that presents itself must be regarded (and treated) as an individual malady that never before occurred in the same manner and under the same circumstance as in the case before us, and will never happen precisely in the same way" ("The Medicine of Experience," *Lesser Writings* 442). Hahnemann used a striking metaphor to illustrate how the infinite variety of diseases in nature could not be arbitrarily formed into classes: "The polyhedrical kaleidoscope held before the eye arranges in one illusory picture a number of external very different objects, but if we look behind it into nature, we discover a great variety of dissimilar elements" ("Representation to a Person High in Authority," *Lesser Writings* 699).

It now becomes clear why Hahnemann recommended the intent listening to the patient, took seemingly disorganized notes in the *Krankenjournale*, as well as accumulated the copious symptoms in the *Reine Arzneimittellehre*: if the manifestations of a malady are in each case different, the diseases infinite in number, and the arrival at a diagnosis of a disease impossible, the oddest symptoms need to be recorded. They, in actuality, became the secret to ascertaining what made the patient unique and distinct. In determining "what kind of symptoms ought chiefly to be regarded in selecting the remedy," Hahnemann thus prescribes that

> we ought to be particularly and almost exclusively attentive to the symptoms that are *striking, singular, extraordinary*, and *peculiar* (characteristic), *for it is to these latter that similar symptoms, from among those created by the medicine, ought to correspond* . . . On the other hand, the more vague and general symptoms, such as loss of appetite, headache, weakness, disturbed sleep, uncomfortableness, &c., merit little attention, because almost all diseases and medicines produce something as general. (*Organon* 173–4, §153)

To restate, for Hahnemann the notion that illness was unique to each patient meant that the physician needed to divine what Benjamin termed the exceptional third element. Only then could the precise remedy that would exactly fit that patient be found.[33]

Specifically, in the search for finding the right homeopathic remedy, in other words, the *Gegenbild* (antitype) that would illicit the same overt symptoms but not be the original disease, the physician needed to read between the lines. For example, these indicators had to appear intermittently in the course of an infirmity. Harris Coulter offers this illustration: "In the treatment of malaria (intermittent fever) Hahnemann notes that the paroxysms of fever (*communia*) are of little use in the selection of the remedy, since they are experienced by everyone. Instead, the physician should look to the patient's symptoms between the seizures of fever (*propria*), since these differ greatly from one patient to the next" (381). Determining the aberrant, random symptoms meant individualizing the patient and establishing a patient profile that was attentive to such things as on which side of the body a pain came, what time of day it occurred, what changes it provoked within the patient's temperament, or what other signs seemingly unrelated to the malady appeared on other parts of the body. Gerhard Bleul puts it succinctly (15–30): what matter are the uniquely characteristic sensations and modalities – the signs – not the main symptom itself.

To give an example of how Hahnemann desired to pay attention to the peculiarity of each symptom, one can turn to how he recommended testers record their medicinal trials. He prescribes that the experimenter

> place himself successively in various postures, and observe the changes that ensue. Thus he will be enabled to examine whether the motion

communicated to the suffering parts by walking up and down the chamber, or in the open air, seated, lying down, or standing, has the effect of augmenting, diminishing, or dissipating the symptom, and if it returns or not upon resuming the original position. He will also perceive whether it changes when he eats or drinks, or by any other condition, when he speaks, coughs, or sneezes, or in any other action of the body whatsoever. He must also observe at what hour of the day or night the symptom more particularly manifests itself. All these details are requisite, in order to discover what is *peculiar and characteristic in each symptom*. (*Organon* 164–5, §133, italics mine)

That Hahnemann was attentive to the defining, peculiar sign does not mean that all the diverse markers did not need to be recorded and taken into account. Indeed, the above passage suggests as much. He specifies that the totality of all the indicators also needed to be addressed, for they too would help select the proper remedy; these could not be ignored. But especially in chronic diseases, beyond the signs that are in the foreground, the equally important singular, symptoms that the patient him- or herself hardly notices any more, the ancillary symptoms ("Nebenzufälle" or "Nebensymptome"; §95), are crucial for deciding upon the cure. Finally, it bears noting that already in the first edition of the *Organon*, Hahnemann stipulates that a person's frame of mind ("Gemütszustand"), in other words, what most idiosyncratically belongs to him or her, can help determine the choice of a remedy. This unique, characteristic, distinguishing element cannot, he says, escape the eye of the perceptive physician (*Organon-Synopse* 192, §211).

If Volker Hess has contextualized homeopathy in terms of eighteenth-century semiotics, then Harald Walach has investigated it in terms of contemporary semiotics, asking the pivotal question: what relations ("Beziehungen") exist between the signs of homeopathic medicines? ("Homöopathie und die moderne Semiotik" 156). In other words, what guarantees that the signs produced by medications, once matched by those provoked by illness, will lead to health? Walach argues that Hahnemann does not speak in terms of informational content or memory effect (in contemporary idiom, molecular or electromagnetic information) being transferred in the remedy to the patient. Instead of such a causal, material, or direct connection, the action of a remedy is spirit-like or dynamic: "Homeopathic remedies are signs, not causes. Their sign character is, however, not fixedly any 'informational' content present in the remedies. It is of a magical nature" ("Magic of Signs" 136). Walach would in principle agree with Hess that "homeopathy can be seen as matching one type of meaning, the one given by the symptoms of the sick person, with another one, given by the symptoms of remedies in the *materia medica*. Homeopathy in fact is applied semiotics" (ibid.). But to say that the "sign character" is "of a magical nature" is definitely not characteristic of Enlightenment rationalism; instead it is Romantic. I would therefore like to supplement Hess's (and my own earlier focus) on eighteenth-century semiotics with Walach's approach. But to do so, I wish to historicize the "magical nature" of the remedy's effectiveness through reference to how the German Romantics conceived analogical inference. Romantic inspirational discernment is constitutionally different from the setting up of parallelisms characteristic of eighteenth-century semiotics.

Around 1800 there is a shift from a regulatory, normative poetics to the belief in idiosyncratic, ingenious interpretation.[34] Romantic reading is, if anything, non-predictable. "There is no universally true kind of reading, in the ordinary sense. Reading is a free operation. No one can prescribe for me how I am to read something or what" (2: 399, #398), writes Novalis. He also exhorted that the true reader must be an extended author (2: 282, #125). Friedrich Schlegel similarly wrote that the true critic is an author to the second power (*Kritische Ausgabe* 18: 106, #927). If the reader is an extended author, then there is no regularization of interpretation. As we have seen, the same is the case in homeopathy, where there is no point in running experiments, as in contemporary pharmaceutical trials, for one can't predict outcomes. The homeopath is as inventive, imaginative, and idiosyncratic as the Romantic reader: both hone in on the odd, dissimilar sign to unify the heterogeneous.

For the German Romantics, *Witz*, with its attention to the unexpected, plays the crucial role in establishing analogies across separation. In his dissertation on the German Romantic literary theory, Walter Benjamin indeed devotes several paragraphs to the Romantic concept of *Witz*, yoking it with mysticism (1.1: 48–9). In both the realization of analogy occurs instantaneously, like lightning ("blitzartig" 1.1: 49). Benjamin cites Novalis: "Wit is the manifestation, an exterior lightning of imagination. Thus . . . the resemblance of wit and mysticism" (qtd. 1.1: 49). *Witz* makes one's style "electric" (1.1: 49). Friedrich Schlegel indeed predates Benjamin in developing a theory of the dissimilar, pointing out that it is the capacity of a "witty mind" to come up with "distant similarities, thereby favoring the absolute unity of the desired mysticism" (*Kritische Ausgabe* 18: 507, #14). A later Romantic, Joseph von Eichendorff (1788–1857) also deploys the image of lightning to characterize how suddenly the imagination associates dissimilarities, "conceiving like lightning the concealed interrelations of the remotest ideas, as if the most extraordinary becomes comprehensible in itself" (4: 334). Similarly linking *Witz* with the ability to discover unexpected relations, Novalis writes: "Wit elicits creativity – it elicits similarities" (2: 649, #732).[35]

The ironic function of *Witz* in yoking discordant minutiae illustrates how Hahnemann makes the principle *similia similibus curentur* work.[36] He does not single out as the most significant symptom the one that is common to all patients but counter-intuitively the opposite: the most dissonant symptom is the solution linking all the conflicting particulars. Incompatibility actually spans the incongruities that the ill body presents. Of course, also odd (*witzig*) as well as uncanny (*unheimlich*) is the fact that a substance producing similar symptoms to the disease actually cures it.

The eminence that the Romantics lent to analogical, inspired thinking helps to explain how *similia similibus curentur* could find the cultural resonance it did in the first half of the nineteenth century. "As principle of relationships," Novalis noted, *Witz* was a "Menstruum universale" (2: 250, #57). In several fragments, Schlegel similarly links *Witz* and analogy with the faculty of extending connections universally: "Wit is universal chemistry" (*Kritische Ausgabe* 18: iv, #440). "The greatest wit would be the true lingua characteristica universalis, and simultaneously the ars combinatoria" (*Kritische Ausgabe* 18: iv, #1030), and "The slightest analogy sheds more light and reveals more spirit of the whole on the universe than travelling to the central sun" (*Kritische Ausgabe* 18: v, # 6). Summarizing the all-encompassing function of analogy, Novalis wrote: "My understanding of the whole would thus have the character of analogy" (2: 340) and "All our sciences are sciences of relationships" (2: 444, #9). He trusted that thinking in analogies and comparisons was at the heart of all knowledge (2: 334, #108). His close friend, the physicist Ritter, believed that life in its essence was "an eternal equation" (*Fragmente* 102, #488).[37]

The Romantic fascination with analogy does not mean, however, that analogical reasoning did not have its detractors at the time. Although Friedrich Schlegel composed an essay, "Lehre von der Analogie" ("Doctrine on Analogy"), on the central importance of analogy in philosophical argument (*Kritische Ausgabe* 13: 314–17), philosophers were among the sharpest critics of its deployment. Hegel, for instance, wrote: "Not only is analogy unfit to provide complete evidence, it rather, by its very nature, refutes itself so often that, to conclude from its own rules, analogy is inconclusive" (3: 193). Goethe, too, recognized the dangers of analogical excess: "If one adheres to analogy too closely, everything will be identical," resulting in a stagnation of observation (HA 12: 368).[38]

But despite this caution voiced by more rigorous, sceptical thinkers, analogical inference broadly characterizes and unites all branches of Romanticism, whether natural philosophy, the life sciences, or poetics.[39] Its roots can be traced back to Herder, who extolled the powers of analogical invention.[40] Following suit, the German Romantics fervently expressed how the signs of nature can be read by uncovering parallelisms or sympathetic attunement. Nature hosted diverse resemblances that promised to manifest a oneness unifying them. For instance, in the illustrious passage at the start of *Die Lehrlinge zu Sais* (*The Apprentices of Sais*) Novalis declares that a delicate writing can be detected everywhere – on birds' wings, eggshells, in clouds, snow, and crystals – a script of ciphers that promises to reveal a magical inner correspondence operative within nature (1: 201). The language of nature itself for Novalis is figural. In reference to the analogical signifying possibilities of the human body, he wrote that it was an "origin of analogy for the universe" (2: 399, #401). Along similar lines, Ritter wrote that one could try to decipher the law of life from the signs it left behind; the art of reading this *Felsenschrift* (cliff ciphers) was far advanced and its ultimate meaning was soon to be revealed (*Physik als Kunst* 45). In the poem "Wünschelrute" (Divining Rod), Eichendorff, too, dreamed of finding the Orphic word that would release the musical harmony within nature: "Schläft ein Lied in allen Dingen, / Die da träumen fort und fort, / Und die Welt hebt an zu singen, / Triffst du nur das Zauberwort" ("A song sleeps in all things / They are forever dreaming, / And the world begins to sing, / Should you only hit upon the magic word"; 1: 112). One of the most famous phrases of the time was Novalis's coinage "Zauberstab der Analogie" ("magic wand of analogy"; 2: 743). For the Romantics, nature led one into its magical forest of echoes, after-effects, and hauntings, its mysterious script ordered by analogies that begged to be read. As Charles Baudelaire (1821–67) wrote in his poem "Les Correspondances": "La nature est un temple où de vivants piliers / Laissent parfois sortir de confuses paroles / L'homme y passe à travers des forêts de symbols / Qui l'observent avec des regards familiers" ("Nature is a temple of living pillars / That at times give voice to confused words. / Man passes by through a forest of symbols / That observe him with intimate glances") (1: 11).

At this juncture, it is crucial to emphasize important distinctions between the Romantic magic wand of analogy and Hahnemann's tenet of *similia similibus curentur*. In the previous section I pointed out how the founder of homeopathy belittled the Paracelsian belief in the signature of things, a belief, by contrast, that Novalis turned into a poetic principle.[41] Hahnemann also does not speculate, as, say, Lorenz Oken does, that the human body and the universe are morphologically analogous: "The universe is nothing but an organism, whose sensorium commune, or self-awareness, is represented by the human body, the brain by the animals, the senses by the plants, while the torso exemplifies the rest that you call inorganic" ("Über das Universum" 10).[42] Novalis, too, theorized that the body "already expresses the analogy with the whole ... The universe is entirely an analogon of the human nature in body, soul and spirit. The former is reduction, the latter is extension of the same substance" (2: 423, #483). He rhapsodized that the whole of nature expresses in its face, gestures, pulse, and colours, the human being: "Does the rock not become curiously familiar when I address it?" (1: 224). In similar fashion, another Romantic, Franz von Baader (1765–1841), trusted that, "just as the comparative anatomy for exterior forms, the sense of analogy of humankind with God and the creatures provides dependable guidance of the self" (12: 173).

But despite Hahnemann's distance from such deliberately poetic and speculative statements prevalent in the Romantic life sciences, he did express a firm conviction in the powers of analogical inference. Specifically, Hahnemann believed that God's creation had hidden within it the cure to all illness – and that he had unlocked the miraculous door via *similia similibus curentur*. His *Materia medica* was a true codex of nature (*Organon* 169, §143). Just as for the Romantics, divine natural order reveals itself in homeopathy through parallel, ultimately cohesive signs. In sum, at the heart of both projects and as their most common denominator lies the desire for unity and connection between man and the natural world. As Peter Sloterdijk summarizes: "Nature herself, with her poetic and guiding qualities, is the heart of duplication and similarities ... There is essentially only one homeopath, nature herself" (19). The structure of analogy confirmed that nature was both law-abiding and organically dynamic.

In conclusion to this chapter, I wish to cite further the address given by the brilliant contemporary German philosopher Peter Sloterdijk on the occasion of the 200th anniversary of the *Organon*'s publication. In it he virtuously juxtaposed the two opposing tendencies in Hahnemann that I have elaborated upon in this chapter: "Everything about him is purely empirical, and everything is infinitely mysterious" (31). He added: "The real and the marvellous coincide" (31). What Sloterdjik does not observe is that this contradiction and concurrence, so productive for the art of homeopathy, arises from Hahnemann exemplifying the two dominant epistemes that he straddled in his lifetime – Enlightenment semiosis and Romantic conjecture.[43] Hahnemann thereby moves from an originally enumerative, observational, empiricist terrain to a fundamentally unmappable, impenetrable, inspirational one. In particular, the Law of Similars, which starts out as a sheer compilation of signs referring to other signs (or symptoms in the body), in end effect requires a honing in on the unusual, exceptional sign – a singling out enabled by sudden illumination or *Witz*. As Walter Benjamin points out, this third term – the moment of dissimilarity – clinches the analogy.

In the third chapter I shall take up again what Walach calls the "magical nature" or the spirit-like actions of the homeopathic remedy that, in tune with the human body, re-energizes it back to health. Then in conclusion I shall develop how Hahnemann's thought resonates with the Romantic desire for harmony with nature via powers that dynamically subtend all of nature. But first, in chapter 2 I look at Hahnemann's foundational belief in the singularity, elevated authority, and authenticity of the individual's experience. It too is a product of its Romantic times.

Chapter Two

THE LAW OF THE SINGLE REMEDY

Singularities: Disease, Patient, Cure

Today many commercial homeopathic preparations take a scattershot approach to treatment. Compilations contain a variety of remedies intended to alleviate related symptoms. For instance, a compound for seasonal allergies with ailments ranging from itchy eyes to sneezing might include homeopathic doses of *euphrasia*, *allium*, and *sabadilla*, as well as *solidago*. A remedy for gastro-intestinal problems from heartburn to flatulence might include *bryonia alba* and *nux vomica*, along with other ingredients. In classical homeopathy, however, only one remedy is administered at any given time, and it is intended to address a plurality of pathological symptoms affecting an entire person from head to toe. Just as each patient is one of a kind, given his or her unique composition of symptoms, so too is the remedy selected unique and not to be adulterated by any other substance. Hahnemann strictly orders in the *Organon* that "in no instance is it requisite to employ more than *one simple* medicinal substance at a time" (218, §272). This, in brief, is the Law of the Single Remedy. He then adds the footnote: "Experiments have been made by some homoeopathists in cases where, imagining that one part of the symptoms of a disease required one remedy, and that another remedy was more suitable to the other part, they have given both remedies at the same time, or nearly so; but I earnestly caution all my adherents against such a hazardous practice, which never will be necessary" (218). Hahnemann indeed excoriated disciples when he learned that they prescribed more than one remedy at a time. He considered it "conformable to nature" to prescribe "one known medicine at a time" rather than "a mixture of several drugs" (218).

In this chapter I wish to build on the discussion in the previous chapter regarding how Hahnemann proceeded in his patient consultations and how he selected a remedy based on the peculiar symptoms of the individual that were dissimilar from other manifestations of what otherwise would seem to be the same ailment. I thus take the Law of the Single Remedy as a gateway to investigate what I consider to be the most modern and novel aspect of homeopathy, namely, its focus on the individual. Because Hahnemann thought that cases of disease do not fit into a universally applicable etiology, homeopathy constructs the unique person as the proper object of medical study. Indeed, he wrote that it was impossible "to obtain a real cure without particularizing each individual case in a rigorous and absolute manner" (*Organon* 144, §82). Karl Rothschuh has summarized that homeopathy went further than all other medical systems of its time in the individualization of the sick person (*Konzepte* 341).

To pursue the implications of the Law of the Single Remedy, I shall rely to a large extent on the extensive, recent transcriptions of Hahnemann's patient notebooks and on the written commentary by their editors. They provide copious, precise, new material for the historical understanding of the patient–physician relation. Above all, they give insight into how Hahnemann recognized the role of the affective life in the practice of the healing arts. Yet, because they have not been translated into English, these texts have remained inaccessible to English-speaking practitioners and followers of homeopathy, a condition this current study wishes to partially remedy.

In addition, I also look at how homeopathy can be situated within the wider medico-historical context. Hahnemann fits within the long tradition of bedside medicine with its focus on the patient's narrative of ailments.[1] Thus, he opposes the standardization of medical practice that was occurring during his lifetime. Hahnemann's emphasis on the health of the whole person places him as well within another major cultural paradigm of his time, namely, the specifically German concept of *Bildung* – the all-round comprehensive education, edification, maturation, and growth of a person. One of the most important aspects of *Bildung* is its stress on the authenticity of one's own unique experience. By 1800 the vehicles of self-investigation through confessions, memoirs, travelogues, and journals had multiplied. They were among the various genres of life-writing that proliferated.[2] In particular, Goethe, Ritter, and Alexander von Humboldt intertwined their autobiographical and scientific writings. The era also saw the birth of the particularly German literary genre of the *Bildungsroman* or novel of personal formation. The fictional character of the *Bildungsroman* cultivated his singular potential through the arts and encounters with social institutions. Its first manifestation and model for subsequent incarnations was Goethe's *Wilhelm Meisters Lehrjahre* (*Wilhelm Meister's Years of Apprenticeship* [1795–6]). Keeping in mind the significance of *Bildung* for this era, I wish to pursue in this chapter the worth Hahnemann lent to the patient's narrative as well as the authenticity and individual experience in his own self-testings. From various perspectives, then, this chapter is devoted to the centrality of the distinctive, individual, intuitive subject for the art of homeopathy – and the importance of homeopathy in the history of the self as broadly outlined by such philosophers and historians as Charles Taylor (*Sources of the Self*) and Jerrold Seigel.[3]

The historian of medicine Roy Porter characterizes the constitutional doctrine of classical medicine as follows: illness

> was an expression of changes, abnormalities, or weaknesses in the whole person; peculiar to the individual, it was "dis-ease" rather than disease. Such a person-centred view could underwrite a certain therapeutic optimism: relief was in the hands of the "whole person." Classical medicine taught that the right frame of mind, composure, control of the passions, and suitable lifestyle, could surmount sickness – indeed, prevent it in the first place: healthy minds would promote healthy bodies. ("What Is Disease" 80)

Although Porter is talking about the Greeks and the learned medicine of the Middle Ages and Renaissance, this prescription for balanced health and mind and emphasis on the "whole person" characterize the neo-Hippocratic philosophies of Hufeland and Hahnemann around 1800. They also are popular concepts in naturopathic and wellness programs today. This long tradition of "a person-centred view" helps to explain why homeopathy gained the foothold it did among the rising bourgeoisie class, despite its oddity and novelty. This tradition also explains the success and longevity of Hufeland's macrobiotics. Homeopathy and macrobiotics are similar in that they link past and present medical methods. They build on a former, well-established doctrine of maintaining a healthy constitution (or, in Hahnemann and Hufeland's words, of a strong vital life force) and on the belief that illness is a result of its imbalance. As well, they lead to the modern conviction that each individual citizen carries the responsibility to tend to his or her own health.

But despite the tradition into which he fits, Hahnemann stylized homeopathy from the start as an alternative, inimitable practice. He denied that it shared features with other medical beliefs, and he was committed to the purism of remaining distinct. For however much homeopathy was devoted to the "whole person," as in Greek medicine, it was also in opposition to the theory of illness dating back to the Ancients that saw people in terms of the four categories of the humours. Although this classificatory Galenist system was long outdated by 1800,[4] its therapeutic procedure of controlling the flows in the body and relieving it of its plethora, as in bloodletting, was still very much in force.[5] Important here is the parallel between these two sorts of singularities: just as homeopathy refused to correspond to medical orthodoxy, it also stressed the distinctiveness of each patient. In focusing on the peculiar symptom, Hahnemann preserved the individuality and autonomy of each person who sought his consultation. What is more, not only is each patient singular, with his or her illness incomparable to another's, but each medicine needed to be unique to that patient. Hahnemann even insisted on the purism of one single remedy to be administered at any given time. In this chapter, I develop the various implications of this philosophy of singularity – both in the theory underpinning homeopathy, as outlined in the *Organon* and in his *Lesser Writings*, and in his well-documented practice of devoting more time to his patients and more note-taking of their conditions than any other medical practitioner of the period.

Homeopathic dedication to the individual is derived in large part from the practice of "biographical" medicine current in Hahnemann's day but nearing its close. John Pickstone describes biographical medicine as "a continuing tradition of seeing illnesses as disturbances of individual lives" (10). This view was superseded by hospital care, in other words, treatment on a mass scale, which had its origins in the collective management of the military's health. Michel Foucault refers to the post-1800 system as the anatomical-clinical method based on the study of cases in institutionalized settings: "a new structure in which the individual in question was not so much a sick person as the endlessly reproducible pathological fact to be found in all patients suffering in a similar way" (*Birth of the Clinic* 119). "The patient has to be enveloped in a collective, homogenous space" (242).[6] As well, the "new, early 19th-century conception of disease as a localized organic lesion, systematically correlated with a group of reported symptoms and observable signs, made the concentration of different doctors on different organs and organ systems a logical turn of events" (Goldstein, *Console and Classify* 60). By the 1830s and 1840s medical specialization began to be widespread.[7] Before the "birth of the clinic," though, the physician was devoted to hearing out all the patient's maladies.[8] To be sure, Hahnemann was not so much a "bedside" physician as a practitioner who, given his renown, saw patients in his own consultation room or received long letters from them minutely detailing their ailments. In fact, he was opposed to doctors making house calls, for he thought it lessened the respect that patients would have for their healers. He did not visit them on their deathbed. Nonetheless, as I shall go into more detail later in this chapter, like the bedside doctor Hahnemann heard out his patients thoroughly.

Therapeutically, the clinic patient was treated and, more importantly, tested on with "specifics." That is to say, rather than the medicine being tailored to the whole person (as in homeopathy) it targeted a specific organ or disease. Andreas-Holger Maehle points out that with the rise of experimental pharmacology, "specific" was defined as a "drug which united its known pharmacological properties (such as astringency, antiseptic power etc.) in such a peculiar and inimitable way, that it was superior to all other drugs sharing those properties" (287). Only with a mass number of patients could clinical trials be widely conducted.[9] Before the rise of institutionalized medicine, however, a specific was a drug which had a precise property such as a diuretic, emetic, emmenagogue, cholagogue, diaphoretic, and so forth.[10] Hahnemann mocked the fact that several herbs could be used for the same purpose: "There arose long lists of simple drugs (*interchangeable remedies, succedaneums*) which were all, without distinction, to be serviceable for one disease" ("Three Current Methods of Treatment," *Lesser Writings* 524).[11] According to him, the dispensing of simples could lead to a poly-pharmacy: with each symptom being targeted by one drug, a patient could end up taking several medications for various symptoms, or even various medications successively for the same symptom.

Hahnemann was equally opposed when drugs were taken together as compounds. Paracelsus, Stahl, and Hoffmann all had their own secret concoctions. Friedrich Hoffmann is still known today for his elixir, "Hoffmann's drops," while *Swedish Bitters* advertise on their label that their tonic formula derives from Paracelsus. Hahnemann attacked apothecaries for increasing "the number of these formulas, for the profit derived from these mixtures was immensely greater than would have been derived from the sale of simple drugs" ("Aesculapius in the Balance," *Lesser Writings* 429; see also "Three Current Methods of Treatment," *Lesser Writings* 524). The other problem with compounds, he noted, was that when mixtures were administered successfully it remained unknown which ingredient was beneficial ("The Sources of the Common Materia Medica," *Lesser Writings* 679). Hufeland was the particular object of this condemnation for having written that "one medicine cannot suffice for all the indications in a disease; several must be given at once, in order to meet the several indications" (qtd. ibid. 666).[12] First and foremost, however, because he believed in the specificity of each individual, Hahnemann rejected compounds. He also rejected specifics by reason of his philosophy that the whole person should be treated.

Feeding into Hahnemann's indebtedness to "bedside" and "biographical" medical practice was his opposition to the medical theories expounded by the Scottish doctor John Brown in his *Elementa medicinae* of 1780, which did not gain ground in Germany until 1797 via the physicians Andreas Röschlaub (1768–1835) and Adalbert Friedrich Marcus (1753–1816), and soon after via Schelling. According to Brown, there were only two types of illness into which a variety of maladies could fit, what he called the asthenic and sthenic; if you will, hypo-stimulated and hyper-stimulated imbalances. Because Brown attributed 97 per cent of diseases to asthenic weakness (Wiesing 68), he largely prescribed only two remedies – opium and brandy – which he regarded as stimulants. Hahnemann considered Brunonian medicine a crude reduction of illness and even more a one-dimensional view of treatment, often requiring copious amounts of medication in order to reverse or palliate the condition. Hahnemann's observations on Brown are recorded in his essays "Fragmentarische Bemerkungen zu Browns *Elements of Medicine*" ("Fragmentary Observations on Brown's *Elements of Medicine*" [1801]), "Aeskulap auf der Wagschale" ("Aesculapius in the Balance" [1805]), and "Monita über die drey gangbaren Heilarten" ("Three Current Methods of Treatment" [1809]).[13] He did praise Brown for dispersing the sway of Galenic medicine, accurately assessing the influence of the six *res non naturales*, refuting the advantage of a vegetarian over a meat diet, and advocating the medicinal effect of a judicious diet ("Three Current Methods of Treatment," *Lesser Writings* 551). Still, Hahnemann excoriated the Scotsman for wishing to "embrace the whole art with a couple of postulates ... to say nothing of the ludicrously lofty, gigantic undertaking of the natural philosophers!" ("Aesculapius in the Balance," *Lesser Writings* 421).[14] Hahnemann condemned the "system-building" of this *naturphilosophische* school of medicine whose "stupendous castles in the air concealed their poverty in the art of healing" (ibid. 422).[15] And he was not alone in his condemnation. Several others, including Hufeland, objected to *Naturphilosophie* as applied to medicine (see Wiesing 144). In addition, Hufeland criticized Brown for not attending to the natural vitalism and self-healing of the body, as well as for not basing his theory on clinical experience. He, like Hahnemann, wished to individualize the patient (Wiesing 78–92), whereas to "German philosophers and physicians, excitability [*Erregbarkeit*] appeared to be [a] dialectic principle, that could explain life and death, health and disease, and also the interaction between organisms and their environment" (Tsouyopoulos, "The Influence of Brown's Ideas in Germany" 66). Indeed, the furore around Brunonian medicine around 1800 helps to explain why Hahnemann's initial writings on *similia similibus curentur* were not more widely received (Schreiber 11). By 1805, however, Brown's influence was in decline in Germany.[16] At this point, Hahnemann could position homeopathy as empirically and therapeutically driven, in contrast to Brunonian schematization based on speculation rather than on bedside experience. This despite homeopathy's own highly wrought system and its similarities to Brunonianism.[17]

Interestingly – for he develops a medical observation into a lyrical insight – the quintessential German Romantic writer, Friedrich von Hardenberg, known as Novalis, also criticized Brunonian medicine for not attending to the individualization of illness in each patient (2: 616, #622), noting that Brown treated the body as a pure abstraction (2: 796, #265) and that his system tended towards the mechanistic (2: 647, #721).[18] The poet of the blue flower wrote, for instance, that every person has their own sicknesses (2: 500, #142; also 2: 828, #390). Regardless of their external environment, everyone's feeling of health, well-being, and contentedness is thoroughly personal (2: 840, #435). "Every individual has his own measurement or relationship to health" (2: 542, #372). In fact, not only was every person unique, Novalis saw the body itself as an unending chain of sheer individuations (2: 796, #265). Herbert Uerlings interprets this passage as saying that the body cannot be seen as a homogeneous entity, something that Brown overlooked: sickness does not arise through quantitative alteration from external stimuli, but primarily from dynamic changes within the organism (172). Whimsically Novalis discerned that most sicknesses tend to be very individual, like a human, or a flower or an animal (2: 797, #268). He continues: "Therefore their natural history, their relations (out of which complications arise), their comparison is so interesting" (2: 797, #268). Noteworthy in this passage are two points. First, despite the individualization of illness, in fact paradoxically because of it, Novalis, like Hahnemann, recognizes the importance of searching out affinity and analogy between disparate entities. For both poet and physician this search was conducted, as discussed in the first chapter, intuitively and idiosyncratically, if you will, poetically. Second, it is not merely that disease and its course are uniquely manifested in each individual but that disease is specific to each individual.

Moreover – and here Novalis goes beyond Hahnemann – because of this specificity, Novalis also conjectured that illness must lead to the development of individuality; it furthered *Bildung*, the German concept of personal education and development. In one of his last fragments before he died of tuberculosis, he mused that especially chronic illnesses offered years of apprenticeship in the arts of living and developing feeling ("Gemütsbildung" [2: 841, #438]). Because of this potential to heighten character, Novalis, in true Romantic fashion, idealized illness over health: "The ideal of perfect health is only scientifically interesting. Sickness belongs to *individualization*" (2: 835, #400). As well, he took the next step and reasoned that if each experience of disease is unique so too must be the cure: he wondered whether there should be several methods of treatment for each disease, just as in music there are several avenues for resolving dissonance (2: 818, #374). In fact, developing the Brunonian notion of over- or under-stimulated bodies, yet contradicting Brown's well-known over-prescription of medication, Novalis contended that the most excitable persons require the minutest stimulation (2: 542, #374). As if anticipating Hahnemann's Law of Minimum, he suggested that all excitable persons receive less (narcotic) medicine. It needed to be extremely diluted and spiritualized (2: 595, #536).

Romanticism vindicated such sensitivity. It defended uniqueness, celebrated individuality, and admired irreducible integrity. A precursor to the Romantics, Johann Gottfried Herder was the first to write in this vein, complaining that the stupidest idea that ever deserved to be thrown into the waste paper basket was that all human souls were the same. No two grains of sand are similar to each other, let alone such powerful abysses as two human souls ("Vom Erkennen und Empfinden der menschlichen Seele" in *Werke* 4: 385). The deepest foundation of our existence is individual, he claimed, as much in our feelings as in our thoughts (ibid. 4: 365). As Charles Taylor put it, for Herder "my humanity is something unique, not equivalent to yours, and this unique quality can only be revealed by my life itself" (*Hegel* 16). And Jerrod Seigel writes that Herder "provided an exemplary model of a peculiarly German way of thinking, which laid great emphasis on the material and relational dimensions of the self while still making every person the project of the self-referential agency that devolves outward forms as reflections of its own inner being" (340).

The German Romantics then elevated this ideal of individuality to new heights. Novalis declared that one needed to absolutize and universalize the individual moment and the individual situation – that was the true essence of romanticization (2: 488, #87). Friedrich Schlegel went so far as to align individuality with divinity: "If every infinite individual is God, so there are as many gods as there are ideals . . . For whomever inner religious service is the goal and occupation of human life, he is a priest. Everyone can and ought to become one" (*Kritische Ausgabe* 2: 242, #406). In turn, Novalis asserted that "every human story should become a Bible – will be a Bible" (2: 556, #433). With every individual's life trajectory being holy, it must therefore be also singular. Novalis thus begins his *Lehrlinge zu Sais* observing that the paths of human beings are diverse (1: 201). In this work he tells the fairy tale of Hyacinth, a youth who finds on his journeys the way back to himself. Upon lifting the veil of the hidden, mysterious goddess of Sais, what is unexpectedly revealed to him is his very own self (2: 374, #250).

With this idealization of the individual in mind, in both literary and medical discourses, one can now interrogate in detail how Hahnemann advanced individual diagnosis and treatment. As noted in the previous chapter, Hahnemann criticized allopathic medicine for attempting to reduce separate manifestations of an illness to one cause, whereas he saw each discrete case as distinctive. He swore that it was always the person with the disease who must be treated, not the disease itself. Hahnemann underscored time and again in his writings the idiosyncrasy and diversity of morbid afflictions. He rhetorically asked: "Can the so remarkably different complaints and sufferings of each single patient indicate anything else than the peculiarity of his disease?" ("Old and New Systems of Medicine," *Lesser Writings* 717). This singularity reflected the "distinct voice of nature" (ibid.). With few exceptions (to which I shall come in a moment), all diseases were "*dissimilar*, and *innumerable*, and so different that each of them occurs scarcely more than once in the world" ("Aesculapius in the Balance," *Lesser Writings* 442).

In fact, each disease seemed to possess its own eccentricity, even personality: the homeopath "judges the prevailing disease according to its peculiarities and phenomena (its individuality) without suffering himself to be led astray to a wrong mode of treatment by any pathological systemic nomenclature" ("Allopathy: A Word of Warning to All Sick Persons," *Lesser Writings* 741). In 1809 Hahnemann averred – in his typically florid style – that "whatever be their outward resemblance – for example the dropsies and tumours, the chronic skin diseases and ulcers, the abnormal fluxes of blood and mucus, the infinite varieties of pains, the hectic fevers, the spasms, the so-called nervous afflictions, &c – present such innumerable differences among themselves in their other symptoms, that every single case of disease must as a general rule be regarded as quite distinct from all the rest, *as a peculiar individuality*" ("Three Current Methods of Treatment," *Lesser Writings* 525).[19] To generalize them into classes "must not only be superfluous but must lead to error" (ibid.). By "error" he meant that physicians would jump to a diagnosis to match the nomenclature they had at their fingertips. They also needed to recall only a couple of prescriptions, as was the case with the Brunonians' dispensing of opium and alcohol. "In consequence of the ease with which conclusions relative to the essential nature of diseases were come to, there could, thank heaven! never be any lack of plans of treatment . . . as long as [the patient's] purse, his patience, or his life lasted" ("Old and New Systems of Medicine," *Lesser Writings* 716).

The implications of the individuality of each ailment are significant in many ways for homeopathic practice. For one, it means that the physician needs to be as "sympathizing and attentive" ("Old and New Systems of Medicine," *Lesser Writings* 718) as possible. Hahnemann fully realized the psychic dimension to the practice of the healing arts. For another, given that any particular drug will have differing effects, Hahnemann recognized the need for relative dosaging, tailored to the patient.[20] For the same reasons, with regards to coffee he noted that the terms "moderate" and "immoderate" consumption are relative: "Each person must fix his own standard according to his peculiar corporeal system" ("On the Effects of Coffee," *Lesser Writings* 394).[21] Hahnemann also maintained that the organism only becomes ill once susceptible; the various causes of disease, even in acute infectious disease, do not always produce illness in everyone or at all times (*Organon* 108, §31).[22] Finally, given homeopathy's comparative system of *similia similibus curentur*, the uniqueness of each disease parallels the varying effect of each medicinal herb on each tester, explaining the vast number of symptoms that are listed for every remedy in the *Reine Arzneimittellehre*.

In fact, only by virtue of each plant, mineral substance, and salt differing in its external and internal qualities as well as its medicinal properties can it promise to match a patient's needs: "Each of the substances effects an alteration in our state of health in a peculiar, determinate manner" ("The Medicine of Experience," *Lesser Writings* 452; see also "On Substitutes for Foreign Drugs," *Lesser Writings* 509). The diversity of the plant world, its "fullness and abundance" testify to "a divinely rich store of curative powers" ("Necessity of a Regeneration of Medicine," *Lesser Writings* 515). The advances in isolating chemical compounds at the time – and hence the birth of pharmacology – meant little to Hahnemann.[23] The chemical agents "tell us nothing of the changes in the sensations of the living man which may be effected by the plant or mineral, each differing from the other in its peculiar invisible, internal, essential nature; and yet, forsooth, the whole healing art depends on this alone. The manifestations of the active spirit of each individual remedial agent . . . can alone inform the physician of . . . its curative power" ("The Sources of the Common Materia Medica," *Lesser Writings* 676). Penned in 1817, this passage shows how the founder of homeopathy links in Romantic fashion the individuality of a plant or mineral to its specific "active spirit."

Given what Hahnemann called the *"inconjungibilia"* of diseases, in other words, his resistance to how nosology classified them into species and subspecies ("The Medicine of Experience," *Lesser Writings* 442), it is clear why, except in two instances, there is a complete lack of case studies authored by him: if each patient is unique, then none can be exemplary. Normally, medicine like law orients a "case" in terms of previous ones. The peculiarity of each homeopathic patient, though, means he or she falls outside the realm of medical knowledge. "Every cured case of disease shews only how that case has been treated" ("Cases Illustrative of Homoeopathic Practice," *Lesser Writings* 767). To boot, in tracking a case history, medicine, again like in law, reconstructs a sequence of events in order to separate essential from non-essential moments. This narrative reconstruction does not occur in homeopathy, where each and every symptom is recorded. Because entries into the *Krankenjournale* appeared by date of consultation not in a dedicated chart for each patient, Hahnemann never reassembled an entire course of treatment. Hahnemann's "case taking" is thus not the same as a "case study or history."

Only after repeated requests did Hahnemann publish two case histories in the first edition of the *Reine Arzneimittellehre* of 1817, only a couple of pages long each.[24] Even here Hahnemann warns, "we can neither enumerate all the possible aggregates of symptoms of all concrete cases of disease, nor indicate *a priori* the homoeopathic medicines for these (*a priori* undefinable) possibilities. For every individual given case (and every case is an individuality, differing from all others) the homoeopathic medical practitioner must himself find them" ("Cases Illustrative of Homoeopathic Practice," *Lesser Writings* 767). What these two published cases intend to illustrate is how the selected remedy covers "most of the symptoms present, especially the most peculiar and characteristic ones" ("Two Cases from Hahnemann's Note Book," *Lesser Writings* 773). Otherwise they report nothing about dosaging, the description of the healing process, or the possibility of a constitutional remedy tailored to suit the personality of the patient.

Although I have stressed Hahnemann's assertion that diseases were "as diverse as the clouds in the sky" ("On the Value of the Speculative Systems of Medicine," *Lesser Writings* 504), late in life, in *Die chronischen Krankheiten*, he did maintain that chronic diseases could be reduced to three evils, all manifested on the skin – sycosis, syphilis, and psora.[25] He considered them to be caused by infectious miasms and the most difficult to treat of all ailments (*Organon* 140–1, §§79–80). In these mature teachings Hahnemann seems to move away from a phenomenological perspective on disease, where he purely observes symptoms, to a causal one. He shifts his focus from the unique, acute case to the long-term, generalizable disease.[26] Yet even here, when he often turned to the same remedy (thuja for sycosis, mercury for syphilis, and sulphur for psora), he believed in the poly-etiology and individuality of illness. For instance, in the *Organon* he recommended that the attending physician inquire into "the particular circumstances in which the patient may be placed in regard to ordinary occupation, mode of life, and domestic situation" (149, §94). Indeed, the homeopath needs to be all the more attentive in listening to chronically ill clients because they "are so accustomed to their long sufferings, that they pay little or no attention to the lesser symptoms which are often very characteristic of the disease, and decisive in regard to the choice of the remedy . . . [They] never entertain a suspicion that there can be any connection between these symptoms and the principal disease" (149–50). In these protracted conditions, the symptoms may seem to have no connection with one another and cover a unique range, yet they all stem from the same latently developing disease. Here too, then, disease is reflected in the totality of symptoms.

The sections of the *Organon* (143–54, §§82–104) that lay out the procedure for the patient interview are the most precise with regard to the chronically ill. Hahnemann recommends that the attending physician listen to the smallest detail (§95), especially because the chronic illness is the most strange and its symptoms can be transitory. The patient's past must therefore be registered thoroughly. The malingering hypochondriac represents a peculiar case: he or she may be overcome by affect and release a delirium of symptoms. Here the very exaggeration and exaltation ("Hochstimmung") of the patient's diction becomes an important symptom in and of itself (§96). By contrast, there may be patients who, out of lethargy, false modesty, timidity, or even stupidity, speak imprecisely or consider their symptoms insignificant (§97). In all cases, Hahnemann recommends that the physician copy down verbatim the patient's wording, remain silent, and desist from interrupting the flow of his or her narrative or asking leading, suggestive questions (§84). As he wrote in 1825, "The disease, being but a peculiar condition, cannot speak, cannot tell its own story; the patient suffering from it can alone render an account of his disease" ("Old and New Systems of Medicine," *Lesser Writings* 718). This attention to linguistic and rhetorical nuance in case taking is perhaps the most telling aspect of how Hahnemann individualizes his patients. The customized anamnesis demands on the part of the physician "an unprejudiced mind, sound understanding, attention and fidelity in observing and tracing the image of the disease" (*Organon* 144, §83). By the same token, he also recommended that the family and friends of the patient be consulted for their observations (*Organon* 194, §§218 and 220).[27] Thus, even though Hahnemann does not locate a disease in a particular organ or trace its origins, still every pain and suffering has a history to narrate. Each individual has his or her life story to tell.[28]

Barbara Duden has noted that the vocabulary for describing the ailing body was richer in the eighteenth century. Unlike today's medical culture, where a specific organ or part of the body is sick, not the whole person, in the eighteenth century a patient would complain of being sick as a whole. Pain was thereby given a personalized meaning (181). Clearly, Hahnemann's entire system of homeopathy is founded on this personalization of the symptom; it explains its success even today in a health-care environment where the allopathic practitioner rarely spends over ten minutes with a client. Hahnemann, by contrast, encouraged his patients to regard the slightest indicator as significant and to communicate it as precisely as possible. One cannot underestimate here the degree of sensuous apprehension and powers of observation required of both patient and physician, as well as the vitality that Hahnemann believed inhered in the slightest ill. Understandably, given the decline of the Galenist framework, the growing failure of Brunonian innovations, and the attribution of disease to uncontrollable nervous conditions, as well as the caustic effects of heroic treatments, the experience of one's body around 1800 must have been exceptionally bewildering.[29] In one sense, by merely listing symptoms and refusing to explain their cause, Hahnemann could have contributed to the disorientation of his patients even more. Yet, at the same time, he also inspires them to regain control of their sickness via the narration of the self – its history, feelings, moods, and pains. He urges self-construction and self-awareness via language, above all at the moment when pain otherwise disables the subject's sovereignty.

A simple *Krankenjournal* entry from 15 to 17 May 1830 illustrates the minutiae of sensation that Hahnemann records. The female patient experiences soon after early rising, but for a short time only, a heaviness of the tongue and a tendency to cry; she is better afterwards, though without crying; menstruation begins in the evening. At noon she notices a skin irritation on her shoulder and part of the arm; before retiring a drawing pain in the foot. Hahnemann goes on to notate when in the following days she feels weak and when perkier – when even she senses a temporary warmth on her cheeks. On 17 May in the afternoon she suffers such pressure and extension in the stomach region that she needs to undress because she cannot stand the slightest sensation of anything on her body that would constrict the stomach. She feels as if the stomach suddenly extends, whereby blood rushes to her face and slowly dissipates. The patient reports this tension occurs frequently with her periods. (See figure 3.)

Recent studies in the history of homeopathy, with their detailed analyses of Hahnemann's patient records and correspondence, confirm this regimen of self-monitoring and self-annotation. Martin Dinges, in analysing patient letters to the founder of homeopathy between 1830 and 1835, summarizes the importance of the homeopathic individualization process: "No other large-scale source offers this kind of detailed self-observation and discursive self-constitution in medical discourse" ("Men's Bodies" 108). Furthermore, "the discourse allows the patient as the discourse object to become a person who has a body and, of course,

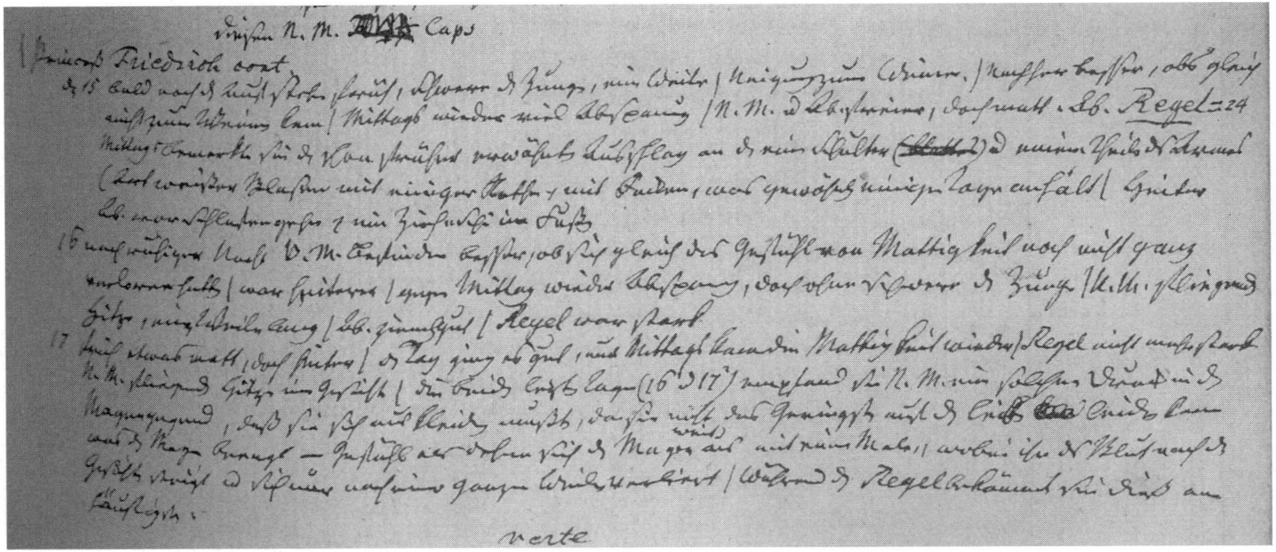

Figure 3 An entry from Hahnemann's *Krankenjournal* of 1830. Courtesy of the Institute for the History of Medicine of the Robert Bosch Foundation, Stuttgart.

also is a body" (107).[30] Other physicians at the time would merely jot down one line or give a Latin diagnosis as a record of the consultation, whereas Hahnemann at the midpoint in his career wrote between ten and thirty lines per patient (Dinges and Holzapfel 150).

Hahnemann also saw fewer patients per day than his contemporaries did.[31] Robert Jütte documents the number and frequency of these consultations:

> When Hahnemann first practised as a homeopath in Eilenburg between 1800 and 1803, he saw 997 patients in 2930 consultations, which makes for an average of three consultations per patient. In Köthen, in the early 1830s, Hahnemann saw (or wrote to) eight patients per day on average... Patients with acute symptoms (such as high temperature) would, in some cases, see Hahnemann as much as three times a day. Long-term patients were asked to present themselves again after seven, fourteen or twenty-one days. For patients who consulted him by letter, the time in between consultations was naturally longer. Up to six months could sometimes pass before the follow-up consultation took place. (*Samuel Hahnemann*, chapter 6, 14–15)

In addition, as Jütte summarizes, "it is a unique feature of homeopathic case histories that they contain more verbatim reports of patients' complaints than any other similar record of medical practice" ("Case Taking" 41).

In addition to one's self-monitoring of the body's slightest ill, Hahnemann stressed the importance of taking responsibility for one's own cure: the patient needed to be active and compliant in his or her own health plan.[32] He insisted not only on detailed communication during the anamnesis but also strict adherence to his prescriptions. Patients were to discontinue consultation with allopaths and the medications they prescribed. The *Organon* as well as Bönninghausen's short textbook on homeopathy were recommended reading.[33] The self-monitoring extended as well to their dietetic regimen.[34] Even in his early writings Hahnemann was against generalized dietetic rules ("The Friend of Health, Part I," *Lesser Writings* 188; "Are the Obstacles to Certainty and Simplicity in Practical Medicine Insurmountable?" *Lesser Writings* 316).[35] In her summary of the case taking from 1830, Ute Fischbach-Sabel similarly comments that Hahnemann's prescriptions vary for diet, depending on the needs of the patient (124–5). On the whole, as Michael Stolberg (*Experiencing Illness* 44) and Jens Busche point out, following dietetic rules gave the populace a sense of self-control, autonomy, and self-determination. People could thereby reject any over-reliance on medicine or superstition. But, as Busche also remarks (35), for homeopathy – as distinct, say, from Hufeland's macrobiotics – it is not a question of "How do I stay healthy?" but how illness can be quickly and safely cured. When Hahnemann prescribed a pure diet, free of coffee and spices, as well as a cleanly environment, free also of fragrances (*Organon* 218, §260), it was so that the homeopathic remedy could perform its desired effects.

A fine example of Hahnemann's prescriptions for a dietetic regimen is his correspondence with the von Kersten family over several years.[36] Jens Busche documents how the head of the household, Friedrich von Kersten, wrote extensive letters to Hahnemann detailing his exercise, the weather conditions, and his appetite, as well as assuring the physician that he was sticking to his diet. Hahnemann wrote affirmative letters in response. This intensive dialogue and affective relation were an important part of homeopathy, presaging, as it were, psychotherapy (Busche 136).[37] Indeed, Princess Luise of Prussia even narrated her dreams to Hahnemann: their correspondence illustrates how Hahnemann lent an open ear – and how effusively appreciative the patient was.[38] It is crucial to recognize that what occurs in all these cases is not just a disciplining of the self, following a tradition of askesis, nor simply discursive self-construction, but a mediatized technology of reproducible selfhood: by registering atomized sensations, plotted consistently into the times of the day, the patient is instrumentalized. He or she becomes a source of registered symptoms – and not just in the *Krankenjournale*: these symptoms migrate as well into Hahnemann's other collections, printed in the *materia medica* and pasted into the repertories.

Of course, a different question is to what did the patients seeking out a homeopath attribute the cause of their illness. Marion Baschin has noted that none of the patients going to consult the homeopath Bönninghausen saw their sickness as arising from religious failings or evil. Some did describe an illness as arising sympathetically with another person or state that, for instance, asthma or incontinence was inherited. Most often, she states, they attributed their ailments to the weather, shocks, strong emotions, food stuffs, or bereavement. None could speak, as one would today, of circulatory problems or of a rising temperature (*Wer lässt sich* 248). Their problems were of a much more fluid nature (249), including weak nerves. Another way of expressing this self-diagnosis is to say that the patients found their problems concerned the whole body and were unique. This judgment corresponds to what Michael Stolberg has generally discerned in patient letters of the period: people opined that their symptoms were absolutely exceptional and that their sufferings were the manifestation of individual distinction ("Patientenbriefe" 30). In this context it is clear how homeopathy, with its theory of individualized illness, could be especially appealing. It moved away from an earlier mechanized, hydraulic view of the body – or even medical explanations based on irritability – to a focus on communicated self-attentiveness.

What, though, do the recently edited *Krankenjournale* tell us about Hahnemann's praxis and the individualization of patient treatment? As to the Law of the Single Remedy, even from the start, he never mixed concoctions; if several remedies were prescribed they were to be taken serially (von Hörsten 65). But, according to Iris von Hörsten's summarizing of the journal from 1801 to 1803, 57.9 per cent of the time Hahnemann changed prescriptions on the next visit, without explanation for the change (71). Likewise, Markus Mortsch observes that in 1821 he switched remedies frequently, every week, in contradistinction to his claim that only one drug could work at a time in the human organism (172). Inge Christine Heinz counted the patient Princess Luise of Prussia receiving thirty-one different medicines out of a total of sixty-six prescriptions between October 1829 and March 1835 (112). Although the princess was an enthusiastic proponent of homeopathy, her own health, as she herself remarked, was in a constant state of change (Heinz 239).

In their analysis, the editors of the *Krankenjournale* have also compiled which remedies Hahnemann prescribed and the frequency with which he did so from patient to patient. Here, too, a marked discrepancy arises. For instance, although he prescribed forty-eight remedies during the period between 1801 and 1803, 56.3 per cent of the time he prescribed only three remedies – *chamomilla, pulsatilla,* and *nux vomica* (von Hörsten 69). By 1817, Hahnemann termed them "polychrests," in other words, special homeopathic remedies that have a wide range of action affecting all tissues in the body. Hahnemann writes in his chapter on *nux vomica*: "There are a few medicines, the majority of whose symptoms correspond in similarity with the symptoms of the commonest and most frequent of human diseases, and hence very often find an efficacious homoeopathic employment. They may be termed *polychrests*" (*Materia Medica Pura*, 2: 223). Similarly to von Hörsten's findings, Reinhard Hickmann has noted that there are several indications that for months at a time Hahnemann prescribed to almost all patients the exact same remedy (418). For instance, in May and June 1824 they received *carbo vegetabilis* (252), while in January 1820 almost all patients were dispatched sulphur (79). By the same token, case taking from 1821 shows that Hahnemann prescribed entire medical sequences from one patient to the next, suggesting that he used them to test remedies (Mortsch 172).[39]

A number of questions arise from these findings. First, the high frequency of modification, as in the case of Princess Luise, raises the question of to what extent the remedies were effective in their cure. Second, the prescription of only a few medicines for several patients hardly rhymes with the central tenet of the *Organon* §153 (as discussed in chapter 1) that the most singular symptom is the key to selecting a remedy. Do then all patients demonstrate the same symptoms? Both the high rate of occurrence with which (1) remedies given one single patient were modified and (2) the same remedy was given simultaneously to several patients raise the question of whether Hahnemann's praxis was truly tailored to the individual.

Hahnemann also breaks with his own theory when he refers to the names of diseases. Between 1801 and 1803, for instance, the most frequently listed illnesses were asthma (20x) and cramps (13x) (von Hörsten 64). But, still, as Ute Fischbach-Sabel observes, for the most part he does not use diagnostic vocabulary. When he does deploy terms such as jaundice or rheumatism, he does not let the name determine the choice of remedy (20). Furthermore, according to Markus Mortsch's list of the most prevalent items for the initial anamnesis in Hahnemann's journal of 1821, diseases are not included. The top eight items are sleep (79.3%), regularity (72.2%), appetite (68.8%), coffee consumption (53%), temperature (52.3%), skin appearance (51.2%), symptoms of the extremities, such as tiredness or falling asleep (49.6%), and mental state (46.6%). According to Marion Baschin, Bönninghausen too would diagnose a specific illness, although, like Hahnemann he also added various other symptoms, which individualized the case (*Wer lässt sich* 264).

In summarizing these various findings from the *Krankenjournale*, one could venture to say that they often raise more questions than provide answers. To be sure, the lack of systematicity can be explained by the fact that these journals were temporary notations, never intended for publication. Insofar as they served as memory aides, Hahnemann did not need to jot down all the symptoms narrated (Inge Christine Heinz

260–1). Thus, only rarely are the questions Hahnemann asked of his patients noted. They do not allow us to reconstruct the state of Hahnemann's knowledge of how effective the remedies were or the reason for their choice. Von Hörsten concludes that the polychrests administered demonstrate no recognizable single defined indication for their dispensation (70). In addition, if, for instance, between 1817 and 1818, 37 per cent of patients do not return after one visit, and 25.6 per cent after two to four visits (Schuricht in the *Krankenjournal D16* 11), we do not know if the reason is that they were disappointed, were cured, or died.[40] As Schuricht points out, we cannot reconstruct clearly the progress of treatment from the notes (ibid. 144).[41] Such issues lead Wischner to conclude that the *Krankenjournale* are more experimental in nature than representative of an application of Hahnemann's teachings (346). But they also raise the question as to what homeopathic teachings were then based on. If Hahnemann did not believe in the general application of a "case study," how could he formulate the laws or principles of homeopathy? This present study wishes to indicate that the scholar needs to search for answers to such conundrums within contemporaneous cultural and medical-anthropological discourses. There is not sufficient evidence to claim that they lie within the results of Hahnemann's medical practice.

Finally, how was Hahnemann's commitment to individual care regarded by his contemporary physicians? The lengthiest response to the *Organon der Heilkunst* was penned by Johann Christian August Heinroth, who wrote an *Anti-Organon* in 1825. Given Hahnemann's own principles and praxis regarding personalized consultation, diagnosis, and treatment, it is unexpected to see Heinroth criticize homeopathy precisely for its lack of individuation – until one recognizes Heinroth's own medical background. His legacy lies in having invented the term "psycho-somatic." Karl Rothschuh categorizes him a *Psychiker* (*Konzepte der Medizin* 315), meaning that he believed mental illness had an exclusively psycho-dynamic root in one's personal environment. Thus, even more than Hahnemann, Heinroth paid attention to a patient's individual life circumstance. He condemns homeopathy precisely for its abstraction (44). He agrees with Hahnemann that there are infinitely diverse manifestations of illness (31–2, 141–2). But he arrives at the opposing conclusion that, if each case is different, one cannot logically compare symptoms (i.e., between the sick and healthy) in order to arrive at the choice of medication. According to Heinroth, then, Hahnemann does not obey his own principles (33). Moreover, Heinroth accuses homeopathy of focusing on an entirety of *symptoms* rather than on the entire *individual* (44) and believing that one can cure with a few drops faults in temperament, a lifetime of indulgence, domestic privation, the influence of economic calamity, and poor prospects for the future (45). By taking the part (symptoms) for the whole (76), Hahnemann fails to take into account the sum of an illness, which must include its development and prognosis. Heinroth adds that Hahnemann bypasses the importance of orthopaedic conditions (35). His polemic, then, interestingly offers a contrastive deployment of the same rhetoric of individuality and wholeness, decidedly core values as well in Romantic literature and thought.

Individuality and *Gemüt*

Notwithstanding Heinroth's assessment that homeopathy does not take into sufficient account failings of temperament, what remains salient about homeopathy, by any other standard, is its attention to *Gemüt*, or one's mental disposition. In fact, by *Die chronischen Krankheiten* Hahnemann was listing the changes in a tester's temperament first and foremost. As previously mentioned, references to a patient's mental state can be found in almost half (46.6%) of the cases in the 1821 notebooks. A salient characterization of temperament can be found in Bönninghausen's portrayal of one of Germany's most famous women writers, Annette von Droste-Hülshoff, who started seeing Hahnemann in 1829.[42] Bönninghausen writes: "A woman in her thirties, blond and with a very excitable disposition and an excellent talent for poetry and music, she had been suffering for some time from tightness of the chest, and had taken it into her head that she contracted consumption from caring for her brother who had died from this disease this past spring" (*Das erste Krankenjournal* 3; see figure 4). *Gemüt* also is the facet of Hahnemann's teachings that later homeopaths,[43] such as James Tyler Kent (1849–1916), Georgos Vithoulkas (1932–), M.L. Sehgal (1929–2002), and Rajan Sankaran (1960–), expanded into the crux of their practice. But, culturally and historically speaking, why did Hahnemann pay more attention to *Gemüt* over and above other symptoms?

From a broadly based perspective, once religious beliefs begin to lose their cultural dominance, the body can no longer be degraded and illness attributed to just retribution for sinful behaviour. The corporeal nature of man, especially in relation to his spiritual and intellectual capacity, becomes open for analysis (Jutta Heinz 21). As well, the

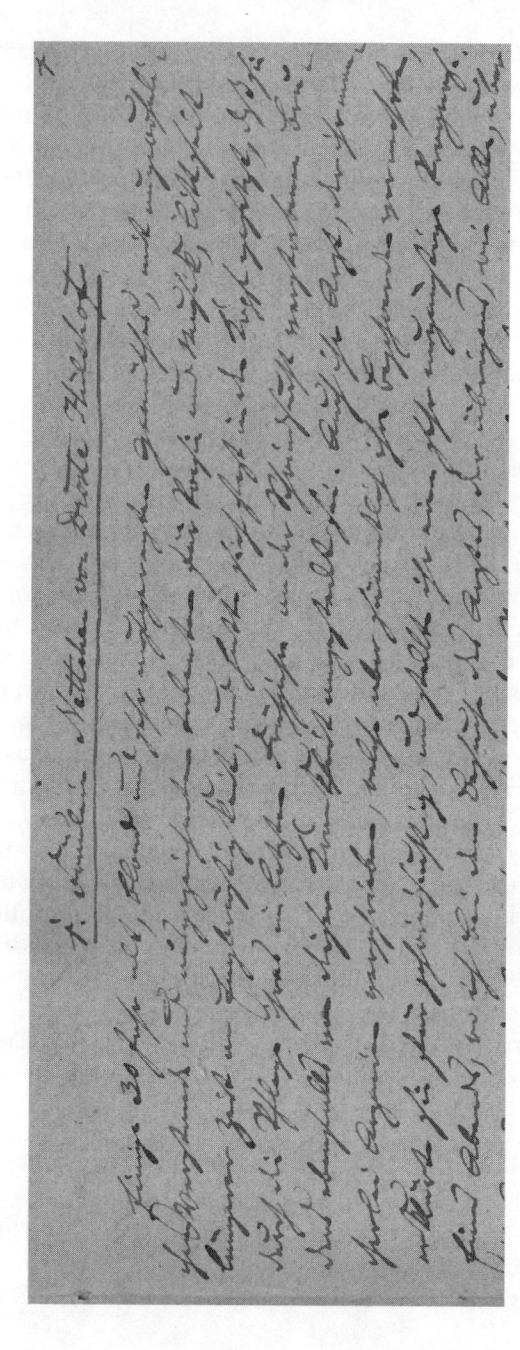

Figure 4 Description of Annette von Droste-Hülshoff from Clemens Bönninghausen, *Homöopathische Heilungs-Versuche*, 1829. Courtesy of the Institute for the History of Medicine of the Robert Bosch Foundation, Stuttgart.

body-mind dualistic legacy of Christianity – exacerbated by the Cartesian division of the two into different substances – begins to lose its sway. At the same time, Julien Offray de La Mettrie's (1709–51) mechanistic, reductive notion of *l'homme machine* was seen as increasingly less credible. Following the eighteenth-century investigations by Stahl, Boerhaave, and Haller (1708–77) into the nervous system, the complex interaction between body and mind became an increasing topic of medical speculation. Stahl, as Rothschuh points out (*Konzepte der Medizin* 305), was important for the movement dedicated to *Gemüt* in the eighteenth century because of his teachings on affects. He maintained that there was no need to know the physiological details of the body, but that the physician needed to read the "world of feeling" (295–6).[44] Haller distinguished between "sensibility" (having to do with the nerves and soul) and irritability (with the contraction of the muscles). A follower of Haller, Cullen coined the term "neurosis" in 1777, though for him it meant physical lesions of the brain. A turning point came in 1772 with Ernst Platner's (1744–1818) *Anthropologie für Aerzte und Weltweise* (*Anthropology for Physicians and Sages*), in which its author maintained that man's harmony between body and soul demanded a new medical discipline.[45] The belief in how the body reflected the soul gave rise to Lavater's physiognomy.[46] As well, popular journals began appearing devoted to *Erfahrungsseelenkunde* (experiential psychology), as is the famous case under the editorship of Goethe's friend and novelist Karl Philipp Moritz (1756–93). Finally, the young Friedrich Schiller (1759–1805) entitled his 1780 dissertation *Versuch über den Zusammenhang der thierischen Natur des Menschen mit seiner geisten* (*Essay on the Unity between the Animalistic and Spiritual Nature of Man*). Here he wrote: "A human being is not composed of a soul and a body, a human being is the most intimate blending of each of these substances" (20: 64).

By 1810, then, in a treatise on contagious diseases, Friedrich Christian Bach (d. 1815) wrote not only that psyche and soma work in tandem in the individual; they can be transferred from one individual to the next (7).[47] In fact, he compared the spread of diseases to the effects of mesmerism, paranormal sympathy, and the transposition of affects (as when everyone starts to laugh at the same time). Further linking body with spirit, he refers to the phrenologist Franz Joseph Gall's (1758–1838) notion that the seat of the reproductive drive lies in the brain (18).[48] By the same token, the Romantic physicist Johann Wilhelm Ritter in 1798 wrote of the "large ribbon" that tied body and soul (*Beweis* 165), allowing him to conduct electrical experiments on muscles and nerves that he believed proved the connection between inner and outer nature. "Nervous" conditions, of course, were fashionable right through the nineteenth century.[49]

What one witnesses, in addition, as one approaches the close of the eighteenth century is a unifying entity taking over the old body-mind (as well as Hallerian irritability-sensibility) dichotomy, whether the *Seelenkraft* as in Herder, the *Lebenskraft* as in Hufeland, or what Manfred Frank has termed *Selbstgefühl*, a self-awareness based neither on the Cartesian cogito nor on mere perception of sensations but on a deep feeling of self. Jean-Jacques Rousseau (1712–78) relishes the experience "of nothing being external to the self and of oneself as God. The sentiment of existence denuded of any other affection is in itself a precious sentiment of contentment and peace" (1047). Denis Diderot (1713–84) writes on *sentiment intime* in his encyclopedia thus: "The intimate sentiment that each one of us has of his own existence and which he senses within himself is the initial source and initial principle of every truth to which we are susceptible. There is nothing more immediate" (15: 57).

Following suite, the German Romantic Novalis asserted the primacy of feeling over reflection (2: 19). The Enlightenment thus ushers in Romanticism. As Jutta Heinz puts it, the dualist mind-body paradigm is superseded by a concept of the individual as diverse being ("individuelles Mischwesen"; 121). She also observes that the Enlightenment problem of the dual nature of man as physical and intellectual gives way to the Romantic focus on the individual. In other words, instead of explaining or describing a phenomenon via the individual, thinkers of the time study individual character (51). Consequently, Romanticism does not make universal, moral claims as did Enlightenment anthropology. Under this new paradigm, everyone's feelings were seen to differ. Over and above this, Novalis stressed the creative, artistic construction of the self: it is "not a product of nature – not nature – not a historical being – an art – a work of art" (2: 485, #76). Life is, in fact, a self-fashioned novel (2: 352, #187). With each person reacting varyingly to the environment, each had a story to tell, whether in the new literary genre of the *Bildungsroman* or in the consultation room of a physician. As we have seen in this chapter, no medical system is as far from making universal, generalizing claims about the human body as homeopathy. As such, it is a product of Romanticism and appeals in its therapeutics to the *Zeitgeist*.

In the largest sense of the word in the nineteenth century, *Gemüt* means the same as soul. It encompasses the entire interior life of a person and is an enlivening, active principle (Scheer 52). Only in the second half of the nineteenth century does the word begin to take on its contemporary connotation of passivity, receptivity, and temporary, fleeting emotions (52–3). *Brockhaus* defines it in 1813 as "the inner principle that stimulates activity in the human being," while *Krug* refers to it in 1832 as an "inner principle . . ., which exquisitely motivates us" (qtd. ibid. 52). In the eighteenth century feelings were associated with the opposite of rationality – that is, carnal desire, oversensitivity, or hypochondria. Karl Rothschuh points out that in the last third of the eighteenth century there are countless writings about keeping the *passiones animae* in check to ensure good health (*Konzepte der Medizin* 309).[50] Autonomous self-regulation was imperative. But after 1800 the definition of feeling shifted and it was seen to reside in and construct one's interiority or inner sense of self. One's feelings, or *Gemütsstimmungen*, were allied with the state of vital awareness (Scheer 53).[51]

It is now clear that, with illness specific to each individual, one's subjective response mattered a great deal to Hahnemann in determining the course of treatment, i.e., the all-important choice of remedy. *Gemütsbewegungen* were not merely temporary, superficial, exterior bodily affects, required to be kept under proper control for healthy living; they were the key to unlocking the uniqueness of each individual.[52] Hahnemann stressed that the state of one's *Gemüt* was often the most decisive factor in selecting the homeopathic remedy (*Organon* 192, 194, and 198, §§211, 220, and 230). Thus, by 1828 in *Die chronischen Krankheiten* he prioritized the psychic symptoms by listing them first. If a patient did not make mention of his state of mind, whims, and powers of concentration ("sein Gemüt, seine Laune, seine Besinnungskraft"), the homeopath was to inquire after them (*Organon* 146, §88).[53] The physician must "likewise endeavour to learn whether the patient's state of mind [Gemüts- und Denkart] is any obstacle to the cure, and whether it be necessary to modify, favour, or direct it" (*Organon* 191, §208). Among the psychic causes of illness that he attended to were unexpected news (*Organon* 107, §29), strong passions (*Organon* 181, §181), "continued grief, anger, injured feelings, or great and repeated occasions of fear and alarm" (*Organon* 196, §225). Hahnemann clearly recognized that emotional processes were registered on the body. Consequently, he would always take complaints seriously, seeing them as spontaneous and authentic communication of distress, rather than – as would readily occur in the eighteenth century with its adherence to normative social behaviour – attributing them to dissimulation, affectation, or oversensitivity, all suggested in the derogatory eighteenth-century German term *Empfindelei*. He also differs significantly from John Brown, who paid attention exclusively to external bodily excitants and stimuli.

As to gauging the improvement of a patient, the psychic symptoms for Hahnemann were also paramount. In "all diseases, particularly those which are acute, the state of mind and general demeanour of the patient are among the first and most certain of the symptoms (which are not perceived by everyone) that announce the beginning of any slight amendment or augmentation of the malady" (*Organon* 210–11, §253). Noteworthy in this passage is Hahnemann's parenthetical suggestion that only the homeopath gifted in the art of perception will notice these slight changes. By improvement he means that "the patient feels more at ease, he is more tranquil, his mind less restrained, his spirits revive, and all his conduct is, so to express it more natural" (*Organon* 211, §253). If the dose is not sufficiently small and "attenuated to the highest degree," moreover, it will plunge "the moral and intellectual faculties [*Geist und Gemüt*] into such disorder that it is impossible to discover quickly any amendment that takes place" (*Organon* 211, §253). Because the changes in the state of mind and disposition were the "principal element of all diseases . . ., there is not a single operative medicine that does not effect a notable change in the temper and manner of thinking of a healthy individual to whom it is administered" (*Organon* 192, §212).[54]

Because *Gemüt* constituted one's innate sense of self, it stood to follow that this inner core would be severely impaired and altered by chronic illness. It is here that Hahnemann advised the closest attention to the affective life, adding paragraphs on it in the fourth edition of the *Organon* in 1829. The complex progression is as follows: what, during the initial stages of a chronic malady, could initially appear as a minor symptom in the disruption of the *Gemüt* develops into the main one, displaces corporeal symptoms, and "subdues their virulence by acting on them as a palliative" (*Organon* 193, §216). Hahnemann then argues that "the disease of the bodily organs . . . has been transported to the almost spiritual organs of the mind [*Geistes- und Gemüthsorgane*], which no anatomist ever could or will be able to reach with his scalpel" (*Organon* 193, §216). In short, Hahnemann reasoned that, in coping with chronic debility, one's very personality is transformed and the newly depressive condition takes precedence even over long-standing somatic impediments. Fortunately, though, the state of *Geist* and *Gemüt* also provided the key to the homeopathic remedy (*Organon* 194,

§220). Indeed, it must if bodily symptoms have been relegated to the background by psychic ones. Hahnemann further argued that these psychic indicators must be treated first, that is, clearly surface to the foreground, before the long-term physical infirmity could even be addressed. Only then could a prolonged treatment of the chronic disease (psora) be undertaken (*Organon* 194–5, §§221–3).

On a related note, Hahnemann maintained that all mental illness stemmed from bodily infirmity and thus needed to be cured in the same manner (*Organon* 192–3, §§214–15). Once other symptoms were treated homeopathically, the psychic instability would disappear as well. On the one hand, it could thereby be argued that he lags behind the innovations of Johann Christian Reil, who began the long tradition throughout the nineteenth and most of the twentieth century (i.e., until the rise of psycho-pharma) of treating mental illness as a separate category. On the other hand, Hahnemann also seems to presage our contemporary view that feeling is a physiological reaction and that psychic health, too, has a physiological foundation. For the psyche to improve, the body must first be addressed (*Organon* 197–8, §229). In other words, he acknowledged the wholeness of the psychophysiological nexus. As well, as was to occur in the development of modern psychiatry, Hahnemann subjected all human emotions and passions to a medical gaze.

Indeed, by virtue of this very attention to the psychic state he parallels Reil, in stark contrast to eighteenth-century approaches. Hufeland, for instance, thought that objective signs of illness were more important than subjective complaints (see Pfeiffer 104). Hahnemann also differs from Kant, Hufeland, Goethe, and Schiller, who advocated a stoic response to illness and a steeling of oneself against the environment. By the same token, he would not attribute the mental diseases of melancholy or hypochondria in eighteenth-century fashion to having too strong, inappropriate emotions. Instead, as seen previously, he let his patients give full vent to their complaints and heard them out. Above all, his attention to *Gemüt* was never to evaluate character morally but to assess disposition. In addition, he acknowledged, rather than dismissed, the powers of the imagination in both sickness and healing (*Organon* 101–2, §17). Reinhard Hickmann, for instance, in his analysis of the case history of Antonie Volkmann during the years 1829 to 1831, notes that even when her symptoms seemed hysteric, Hahnemann took her seriously and did not judge her.

In short, in Hahnemann one encounters the fascinating paradox that, although he rejected knowing what transpired physiologically within the body, he read the innermost part of the body – the *Gemüt* – and foregrounded it. In the first instance, he scorned pathological explanations for being too systematic; the body was expressive, not hydraulic. In the second instance, the *Gemüt* gained in importance precisely because it was not mechanistic; it revealed the true, unfolding self and its imbalance in disease. If we now situate Hahnemann in terms of his contemporary Johann Christian Reil,[55] we see both refusing to pathologize mental illness due to moral failings. Both moved away as well from the materialism of iatrochemical and iatromechanical explanations of disease. In his *Rhapsodieen*, Reil classifies three means of healing: the chemical (including dietetics, pharmacology, and toxicology), the physical-mechanical (including chirurgy), and the psychic. Although Reil did acknowledge that "disorders of the soul cause physical illness, physical illness causes disorders of the soul" (*Rhapsodieen* 40),[56] his emphasis was on developing the field of psychic treatment.

But insofar as Reil's psychotherapy often involved painful bodily stimulus,[57] he parts way with Hahnemann. Reil was not alone here. In 1804 Giovanni Aldini (1762–1834) started using electric shocks. Other physicians in the first half of the nineteenth century, including Heinroth, Karl Friedrich Burdach (1776–1846), Karl Georg Neumann (1774–1850), Johannes Müller (1801–58), and Johann Dietrich Brandis (1762–1846), followed suit in believing that prodding the body through painful intervention would bring one back to a consciousness of self. These physicians advocated restoring body-soul integration by forceful means that also included cold baths and bloodletting. Although Pinel had released mentally ill patients from their chains, these other means of treatment seem just as harsh as imprisonment. Roland Borgards explains the reasoning behind such therapy: "When life is at danger of slipping away, it can be held down with pain; when life takes the wrong direction, it can be guided back onto the right path with pain" (435). Pain is therefore the precondition, not the product of the self (436). Homeopathy, by contrast, argues that if the body is correctly treated gently, the mind will improve along with it. Indeed, the healing body needed to avoid even all dietary intoxicants as well as physical and mental exertion for the homeopathic stimulus to work. As to the use of corporeal punishment in mental institutions, Hahnemann interpreted it as the venting of frustration when other treatment methods fail (*Organon* 196–7, §228).

What, though, did Hahnemann write about mental illness? As long as insanity was not full blown, comforting or reasoning with the patient would help. But if the mental instability truly stemmed from a corporeal source, this well-meaning verbal communication, because it acts merely palliatively, would aggravate symptoms of depression or madness (*Organon* 195–6, §224). The body resists and reacts too strongly. In such cases, Hahnemann underscores that the patient must never be excoriated. Physicians and nurses must always maintain the appearance of respecting the patient's reason. Opposition as well as timidity and surrender were counter-indicated (§229). Hahnemann recommended further: "To the furious maniac, we are to oppose tranquillity and unshaken firmness, free from fear; to the patient who vents his sufferings in grief and lamentation, silent pity . . .; to senseless prattle, a silence not wholly inattentive; to discussing a detestable demeanor . . . entire inattention" (196–7, §228). He thus resembled Pinel in that he engaged the patient's intellect and emotions and advocated *remèdes moraux* instead of repressive measures.[58] But, in end effect, the treatment relied primarily on homeopathic preparations.

To set Hahnemann apart from Reil and other psycho-dynamic medical practitioners at the time, however, is not to say that for him,[59] too, pain does not constitute the individual. In fact, homeopathic recognition of the psychic reality of pain was revealed, as we have seen, in its unusually detailed narration and registration during the anamnesis. Ute Fischbach-Sabel comments that from the early to the late journal notations symptoms of pain are depicted more thoroughly and occur more often than any other symptoms (20). These case journals as well as Hahnemann's *materia medica* are replete with precise, rich vocabulary for describing pain. Thus, Borgards's characterization of the new discourse on pain applies equally to homeopathy: "Around the year 1800 pain was not an exterior threat for the human being, but rather a struggle from the inside out; for the physician pain was an enemy as well as the reason of life" (123). Moreover, "as late as around 1800, a narrower approach to the definition of pain and sense of self was established" (152).[60] Whereas today pain is masked and dimmed by analgesics and anaesthetics, pain in the homeopathic anamnesis is quintessential for individual expression. But whereas Reil and Burdach saw physical pain as a stimulus to prompt the mentally ill patient back to a sense of self, Hahnemann maintained that the counter-irritant, in the form of the homeopathic remedy, need only be slight for the mind and body to regain balance.

The Authenticity of Experience: Hahnemann's Self-Testings

As mentioned in the Introduction, Hahnemann published his own compendia of medicinal substances. His first homeopathic pharmacopoeia was the *Fragmenta de viribus medicamentorum* (1805), in which twenty-seven remedies are listed, forerunner to the German-language six-volume *Reine Arzneimittellehre* (first edition 1811–21; second edition 1824–7), listing sixty-three remedies. His last major work, the five volumes (beginning in 1828 with volume one) of *Die chronischen Krankheiten*, is also primarily a collection of homeopathic remedies. The term *reine* or, in the English translation, *pura*, refers to the pure symptoms of a drug, that is, as produced on a healthy individual not affected by any symptoms of disease (*Organon* 154–5, §108). In his essay "Beleuchtung der Quellen der gewöhnlichen Materia medica" ("The Sources of the Common Materia Medica"), which accompanied the 1817 edition of *Reine Arzneimittellehre*, Hahnemann pronounced that homeopathy "administers no medicines to combat the diseases of mankind *before* testing their pure effects: that is, observing what changes they can produce in the health of a healthy man" ("The Sources of the Common Materia Medica," *Lesser Writings* 694). Hahnemann's *materia medica* thus differs substantially from others in that it lists the effects of substances not their medicinal purpose. But why test on the healthy individual and, above all, on the self? In the 1819 edition of the *Organon*, in a footnote to §152 (*Organon-Synopse* 558), Hahnemann mentions that up until six years ago he had only conducted provings on himself. What is the status with which this self is endowed? Why not accept the authority of years of tradition about the effects of substances and their levels of toxicity?

Hahnemann chastised the standard *materia medica* for basing their findings on the prevalent practice of dispensing compounds, which he felt would never allow one to see the true effect of any one medicinal substance. He was equally against determining the medicinal effects of drugs by their smell or taste, especially the aromatics, which were pronounced to be excitants or "strengtheners of the nerves" ("The Sources of the Common Materia Medica," *Lesser Writings* 672). Smell and taste do not tell us anything about the "most important of all secrets, the internal immaterial power possessed by natural substances to alter the health of human beings" (ibid. 672). Hahnemann also rejected the findings of chemistry, for they would never be able to extract how a plant acts "dynamically on the susceptible spiritual-animal organism, in a spiritual manner" (675). Just as each individual person is unique, so each plant or mineral differs "from the other in its peculiar invisible, internal, essential nature" (677). Each individual remedial agent manifests an "active spirit" (672).

In §§121–45 of the *Organon*, Hahnemann outlined his directions for how to test remedies on the healthy. He himself had a store of collaborators on whom he felt he could rely. Section 126 stipulated that these testers needed to possess a conscientious, credible, and scrupulous character. As all medicinal agents produced variable changes in a living organism, all symptoms would not appear in any one individual and not at the same time of day as in another (§134). In order to ensure a purity of response and to combat this diversity, here too Hahnemann prescribed mono-medication and a diet utterly free of stimulants for his testers. Yet, only when observations had been repeated by a great number of testers could he acquire an accurate knowledge of all the effects a medicine was capable of producing (§135). Still, the experiments that the physician made on his own person were preferable to a store of test data (§141). Hahnemann writes, "A thing is never more certain than what it has been tried on ourselves" (§141). He adds in a note that these experiments taught homeopaths "to understand our own sensations, minds, and disposition, which is the source of all true wisdom" (§141). Finally, he saw that "a materia medica of this nature shall be free from all conjecture, fiction, or gratuitous assertion – it shall contain nothing but the pure language of nature, the results of a careful and faithful research [*reine Sprache der sorgfältig und redlich befragten Natur*]" (§144). How difficult this procedure was for his testers is indicated by one of his students, Franz Hartmann, who remarked that special attention was required to notice the imperceptibly discernible symptoms, that is, the most important, peculiar, and characteristic ones, of far superior significance than those that appeared tumultuously (Haehl 2: 100).

Writing in 1850, Hartmann provides further insight into how these provings were to be conducted. Among the substances to be avoided were "coffee, tea, wine, brandy . . . as well as spices, such as pepper, ginger, and also strongly salted foods and acids. [Hahnemann] did not forbid the use of light white and brown Leipsic beer. He cautioned us against close and continued application to study, or reading novels as well as against . . . cards, chess, or billiards . . . [He] advised moderate labour only, agreeable conversation, with walking in the open air, temperance in eating and drinking, early rising; for a bed he recommended a mattress with light covering." The collaborators carried a tablet and pencil with them at all times in order to jot down every sensation as it occurred, specifying the time of occurrence. Drops of the vegetable essence or tincture were to be mixed "with a great quantity of water, that it might come in contact with a greater surface than would be possible with an undiluted drug." They were to be taken in the morning upon fasting. If no effect was noticed, within three to four hours a few more drops could be taken, after which Hahnemann "concluded that the organism was not susceptible to this agent" (Haehl 2: 100). Hartmann adds that "if after the first dose only faint symptoms made themselves felt, I could rely with certainty that with every hour they would become more prominently developed and more characteristic" (Haehl 2: 101). He discovered that, if by trying to accentuate the clarity of a symptom, he tried a second more powerful dose, no further symptoms occurred. Hartmann openly states that the collaborators knew the drugs they were testing; in other words, they were not blind trials – a concept that would have been foreign to him, in any case, at that time.

Hartmann points out that, in order to avoid toxicity, Hahnemann determined for his testers in advance the number of drops to be taken. Having previously proved for the most part these substances on himself or his family, Hahnemann was "sufficiently acquainted with their strength and properties" (Haehl 2: 100). Curiously, though, there is little archival evidence of such self-testing. Contrary to these extensive recommendations based on his own experience, the slim protocol books (G2 and G3) housed in the IGM contain far fewer examples of self-testing than they do notes compiled from various authors, such as Cullen, on the medicinal effects of a substance. Investigating the sources of the *Reine Arzneimittelehre*, Lucae and Wischner ("Rein oder nicht rein?") have observed that by no means is it solely a collection of provings on the healthy; it contains references to several sources, including patients' symptoms. Lucae and Wischner excuse this contradiction of homeopathic principle, because, they say, the *Reine Arzneimittelehre* needed to be assembled quickly. When practice thus contradicts theory, though, homeopathy appears more and more as a purely formal, even poetic edifice. In terms of a history of rhetoric in science, why would Hahnemann have wanted to appeal to the authority of self-experimentation?

Hahnemann was not the first to recommend self-administration of drugs.[61] It has been pointed out that Cullen and Störck were models for Hahnemann in this respect.[62] The difference, however, is that his predecessors were testing drugs on themselves in order to ascertain on behalf of their patients a safe dosage, not out of a conviction that self-testing was the most authoritative technique. In a sense, then, homeopathy actually aims for the opposite goal: self-testing is conducted not to determine safety but to harness and instrumentalize toxicity. In his work on eighteenth-century pharmacology, Andreas-Holger Maehle determines that the "'case history approach' was the *contemporary* method to ascertain and evaluate therapeutic efficacy" (5). But whereas a high percentage of articles on *materia medica*, pharmacotherapy, and poisons in eighteenth-century British journals were devoted to patient case histories, few if any report self-experimentation (Maehle 31–4). The above paragraphs from the *Organon* indicate, by contrast, the store that Hahnemann placed in his own experience because of its authenticity. He involved his own person because he felt that it offered a more reliable, embodied mode of knowledge. Because he considered the inner functioning of the body to be unknowable, Hahnemann needed to provoke the body into legibility, even if it meant taking the step of simulating the ill body on his own person. Paradoxically, then, simulation was all the more accurate because it was personalized. As the direct medium of nature (§144), Hahnemann does not express any worry about what we would refer to today as subjective distortions, biases, or wayward judgment. After all, according to homeopathy symptoms manifest themselves in any case individually from person to person: every disease is idiosyncratic. The subject's own sensory registering, then, becomes the guarantor and purveyor of truth.

Boerhaave formulated the concept of rational empiricism as consisting of observation and description. Generally, Hahnemann is seen as indebted to this eighteenth-century empiricist belief in the reliability of precise, documented observation and to Kantian reason grounded in the enlightened subject.[63] Accordingly, the internal workings of nature are revealed not so much by the grace of God as through the human sensory organs. But this view that positions Hahnemann exclusively as an Enlightenment thinker only goes half way. It sets up a binary opposition between Enlightenment and Romanticism, between observation and speculation, and, tacitly, between good and bad, that all too conveniently represents homeopathy as a scientifically verifiable practice acceptable for the twenty-first century. It thereby does not do justice to the methodological and philosophical complexity of *Erfahrungswissenschaft* (science of experience) at the turn of the nineteenth century.[64] Empirical accuracy for thinkers from Goethe through Ritter to Alexander von Humboldt does not exclude the interpretive field, active participation, the rich teachings of experience, and intuitive as well as aesthetic judgment. Another way to put the issue at hand is to recognize that Hahnemann takes eighteenth-century empiricism to the limit where it flips into its opposite, Romantic science. Because Kant rendered recognition of the exterior world dependent on the perceiving subject, he shifted perspective from the former to the latter. The Kantian turn then led to Fichtean philosophy with its focus on the ego, as well as to the Romantic self-experimentation of Hahnemann, Humboldt, and Ritter.[65] All three scientists were not merely observing and taking protocol; they inserted their own body into the experiment. To be sure, the results are based on empirical, sensory perceptions, but paradoxically any distortion of these perceptions (introduced either by the substance being consumed, the electric shocks administered, or subjective bias), far from being suspect, actually guaranteed the honesty and accuracy of the experience. As I shall elaborate and clarify in the following paragraphs, this shift in emphasis around 1800 is subtle but crucial.

Two scholars, Katrin Solhdju and Jürgen Daiber, have written extensively on nineteenth-century self-experimentation, contrasting it with eighteenth-century tenets. Solhdju points out that the method of self-observation stood under suspicion as unscientific by Kant (13). She elaborates: "Precisely because the mind, needing to be observed, is subject to alterations (or falsifications) due to the activity of observing, the possibility of an objective, calculable, and thus scientific discovery is with respect to such phenomena, according to Kant, *a priori* impossible" (14).[66] Daiber isolates three criteria for the methodically complete experiment in the wake of Newtonian physics: the experiment needs to be calculable (*reductio*), the effect appears repeatedly in the exact same experiment (*compositio*), and the results cannot go beyond the bounds of the empirical grasp (*resolutio*) ("Selbstexperimentation" 50). All three of these rules are broken by Hahnemann. For one, his testings were "calculable" only in so far as they were accumulative registers of corporeal symptoms, but they were always based on personal monitoring and presumed their reliability. For another, even though various testers were used, in principle, given homeopathy's attention to individual response, Hahnemann was not concerned with replicating the conditions of each experiment. Finally, and most importantly, late in life Hahnemann was carrying out trials not with the original tincture of an herb but with highly diluted doses that he claimed more powerful because spiritualized.

Instead of these Enlightenment criteria, one finds in Hahnemann what Katrin Solhdju distils as the three topoi of nineteenth-century self-experimentation:[67] one, the fantasy of reducing the difference between subject and object; two, the notion that one could let nature speak for itself, that is, be able to grasp directly her primary qualities; yet also, three, the idea of making visible or palpable previously invisible realities (8). By becoming the object of his own experiments, Hahnemann aimed to narrow the gap between subject and object. As noted previously, he believed that he was bringing to voice the "pure language of a carefully and conscientiously studied nature" (*Organon* 169, §144).

Most importantly in terms of Romanticism, he trusted he was uncovering through an analogy of symptoms the healing force inherent in a nature otherwise concealed.

In his book on historical epistemology, Hans-Jörg Rheinberger states that the more intimate a scientist is with his experimental procedure the more effective its inherent possibilities become. The more tightly an experimental system is linked to the talent and experience of the researcher, the more independent it makes itself in his hand (*Historische Epistemologie* 42). This virtuoso performance (Rheinberger, *Experiment* 21) is, notwithstanding, a dangerous tightrope to walk, not least because of the dangers of self-toxicity in Hahnemann's case or electrocution in Ritter's. As well, the methodological peril, at least from a contemporary perspective, is that, as Solhdju highlights, inner experience and with it the subjectivity of the experimenter supersedes the exact reproducibility of the experiment in epistemological interest (12). Hahnemann would not have recognized the lack of impartiality in his observations. Moreover, the individual principle operative in homeopathy would prohibit the applicability of double-bind studies. The ideal of the self-denying scientist aiming at impartial objectivity is instead a product of the mid-nineteenth century. Only after Auguste Comte (1798–1857) does positivism take hold, that is, the concession that subjectivism needs to be separate from the object of the experiment. As Lorraine Daston and Peter Galison in their book *Objectivity* point out, it was only at this later date that "men of science began to fret openly about a new kind of obstacle to knowledge: themselves" (34). For Kant and his contemporaries, by contrast, "'objective validity' (*objektive Gültigkeit*) referred not to external objects (*Gegenstände*) but to the forms of sensibility (time, space, causality) that are the preconditions of experience" (30). In sum, then, with the lines between objectivity and subjectivity not so strictly drawn, subjectivity was considered an essential part of the human condition, especially in the pursuit of knowledge.[68]

It is not out of the ordinary, then, that, in order to illustrate the keen, perceiving eye of the physician, Hahnemann turns to the aesthetic realm – to the metaphors of the naturalist illustrator and portrait painter in an essay entitled "Der ärztliche Beobachter" ("The Medical Observer") that appeared in the 1817–18 edition of the *Reine Arzneimittellehre*. He suggests that to educate the patience of the physician it is useful to study the art of drawing from nature, "as it sharpens and practices our eye, and thereby also our other senses, teaching us to form a true conception of objects, and to represent what we observe, truly and purely, without any addition from the fancy" ("The Medical Observer," *Lesser Writings* 725). As well, the physician needed to be like the portrait painter who would pay "attention to the marked peculiarity in the features of the person he wished to make a likeness of" (ibid. 726). Further, in the ekphrasis of his note taking, the physician must be aware that "a single word or a general expression is totally inadequate to describe the morbid sensations and symptoms, which are often of such a complex character, if we wish to portray really, truly, and perfectly the alterations in the health we meet with" (ibid.). Noteworthy in these passages is Hahnemann's reflection on the importance yet difficulty of linguistic representation, aligning it with the precise representation that is required in the fine arts. Daston and Galison discuss how eighteenth-century naturalists worked together with draftsmen and engravers in order to render individuated detail with a calibrated eye and hand. They argue that the concern during this period is for accurate, painstaking fidelity to the image rather than for objectivity, as we understand the concept today. They point out that "seeing – and, above all, drawing – was simultaneously an act of aesthetic appreciation, selection, and accentuation" (104).[69] Hahnemann similarly writes in his important article "Heilkunde der Erfahrung" that all the physician "needs is carefulness in observing and fidelity in copying" ("The Medicine of Experience," *Lesser Writings* 443), to which he appends a note on how much time it requires to draw a single striking portrait. Again he uses the metaphor of the "accurate picture" of the disease (ibid.).

Hahnemann stands with other several important thinkers of his time who theorized the necessity of personal wisdom, sensory involvement, and intuitive judgment in scientific research. Like Goethe and Humboldt, Hahnemann was dedicated to an active beholding, personal collecting, and attentive describing of phenomena.[70] In response to Schelling, for instance, Goethe said in a letter to Schiller dated 21 February 1798, "I cannot concur with an idea that forces me to waste my collection of phenomena" (WA 4.13: 77). Hahnemann, too, objected to philosophical and theological systems. He was part of the eighteenth-century collection mania of observable data; but he was also, like Goethe, interested in life itself. Like many intellectuals of the day, he was searching for scientific proof for the oneness of nature. What binds Hahnemann to his contemporaries Novalis, Ritter, and Goethe, then, is the desire to find commonality among disparate entities. They longed to see the unity and connections between man and the natural world. What Astride Tantillo writes of Goethe could thus equally apply to Hahnemann: "He focused his inquiries in those areas outside of a mechanistic program, such as an organism's striving to overcome obstacles and limits, the dynamic relationship of an organism's parts that could be understood only in the context of the whole organism, and in an organism's active relationship to its environment" (194).

Most importantly, all these thinkers realized that they could not separate objective observation from intuitive appraisal, subjective registering, affective involvement, aesthetic judgment, and creative imagination. For them, the observation, collection, and presentation of knowledge were justified by subjective evaluation. Knowledge was demarcated by point of view. In 1808, for instance, Goethe wrote to Carl Friedrich Zelter (1758–1832) that the body as an instrument of measurement was the most precise physical apparatus in existence (WA 4.20: 90). Most famously, he formulated the notion of a tender or delicate empiricism that demanded of observation identification as intimate as possible with the object of inquiry: "There is a delicate empiricism seeking to become one with the object and thus becoming the actual theory. But this heightening of mental powers belongs to a highly educated age" (HA 12: 435, #509). For Goethe, then, this subjective identification did not entail a lack of precision; on the contrary, it led to the theorization and elevation of the process of investigation. As Bernhard Kuhn writes, "Objective, empirical observation blends seamlessly with subjective impressions. To separate the two becomes impossible . . . That is exactly the point . . . The difficulty of clearly demarcating the boundary between self-expression and the exploration of nature testifies to the deep-seated interrelation between nature and human nature that lies at the heart of the project of natural history in the romantic period and of the writings of Rousseau, Goethe, and Thoreau" (16). Tantillo similarly summarizes: "Whereas Descartes advocated a complete separation between the thinker and the world, Goethe focused upon the individual's relationship with the world. He argued that it is impossible to conduct truly objective experiments because all scientists, whether they admit it or not, theorize each time they examine the world" (3).

The prime example in Goethe's scientific writings of this theorization is his notion of an archetypal plant or *Urpflanze*, both the basis and idealization of all actual plant species. This extrapolation was premised on his empirical observations of plant morphology, but it was nonetheless a radical reassessment of empiricism, which legendarily led Schiller on 20 July 1794, after a Sunday meeting of the Society of Natural

Science, to call Goethe's *Erfahrung* an *Idee*, which could be translated as a concept of reason. Jürgen Daiber concurs, noting that, when Goethe derives from his observation of empirical individual cases the pure phenomenon or *Urphänomen*, the latter "refers to an invisible, ideal world beyond the individual phenomenon" (*Experimentalphysik des Geistes* 131).[71] Finding more of a middle ground between Goethe and Schiller, between *Erfahrung* and *Idee*, Bernhard Kuhn again writes: "The archetype is not merely a creation of the subjective consciousness, a figment of the creative imagination, or an *Idee* as Schiller would understand it; nor is it a single variable and isolable phenomenological fact that can be pinpointed to in nature. Goethe's scientific writings constantly attempt instead to negotiate the space between subject and object wherein the archetype reveals itself" (70–1).

Despite this idealization, as Hegel (1770–1831) recognized, Goethe had one foot rooted solidly in empirical, sensory inspection of nature, especially when compared to Schelling and his followers. In a letter to Goethe dated 24 February 1821, Hegel noted the poet's sensory and judicious observation of nature ("sinnige Naturbetrachtung") in opposition to the purely intellectual discernment ("begreifenden Erkennen") of *Naturphilosophie* (*Briefe von und an Hegel* 250). Indeed, Goethe was fully aware of the dark spectre of conjecture that loomed large in Romanticism. That is to say, at the same time that he advocated for the role of intuitive judgment in scientific investigation, Goethe also spotted the limitations of self-experimentation, as if stepping back from the precipice of the invisible. He cautioned, for instance, in his *Maxims and Reflections* that there was always a blind spot in one's own vision that one cannot see: "When a human being pays particular attention to this spot, he falls deep into a mental illness, evokes illusions from his other world that are in truth delusions and that have neither form nor limit, but rather frighten as hollow night terrors, haunting in the most ghostly manner whomever cannot break away from them" (HA 12: 373). Another writer of the period, Heinrich von Kleist (1777–1811), recognized the blindness that came from a subjective coloration of the world. He reckoned what would happen if all human beings had green glasses instead of eyes: they would never be able to discern whether they saw things as they are, or whether something was added that did not belong to them (2: 634). Kleist's pessimistic point is that we all have a subjective perception of the world, but we don't realize our limitations.

The Romantics Novalis and Ritter are no less important than Goethe for their pronouncements on the necessary self-involvement of the researcher, but they move much more into the speculative, otherworldly, psychic realm from which he recoiled. Jürgen Daiber has investigated in various publications how for the Romantics empirical verification means testing on the experience of the subject, mentioning Achim von Arnim (1781–1831), Alexander von Humboldt (*Experimentalphysik des Geistes* 106), Ritter, and Novalis. Novalis, for instance, wrote about how experimentation exercises self-observation: "From experimenting we learn to observe – by experimenting we observe ourselves etc. – and by doing so we learn to draw from the strange phenomena reliable conclusions about their unity – that is to say, to observe closely" (2: 657, #766).[72] Furthermore in a fragment on medicine he muses, "No observation without reflection – and vice versa" (2: 567, #451). "The process of observation is simultaneously a subjective and an objective process – an ideal as well as a real experiment" (2: 594). According to Novalis's epistemology, "selfhood is the basis of all knowledge" (2: 670, #820). In particular, he addressed how for his friend Ritter the result of every experiment is bound up with the "necessity of an assumption of the individual, as the final motivation" (2: 816, #368).

It was noted above how Hahnemann used the metaphor of the illustrative artist to exemplify the calling of the physician to precision and exactitude. In reference to Ritter's experiments, Novalis similarly opined that the true observer of experiments is a gifted, genial artist: "Experimentation implies natural genius, i.e., it is the marvellous ability to sense the essence of nature – and to act in harmony with it. The real observer is an artist – who discerns the significant and knows how to identify the important matter from the strange and transient mixture of phenomena" (2: 471). Goethe, too, perceptively saw how Ritter's empiricism was inseparable from his self-testing, but unlike Novalis's awe, it evoked in him horror. Goethe wrote that, in the physicist's presence, he felt as if the evil angel of empiricism was hitting him with his fists (WA 4.13: 204).[73] Ritter's strictly inductive protocol of self-administering shocks from the Voltaic column in order to accurately observe them appeared gruesome to Goethe, and understandably so: it led to the premature deterioration of Ritter's health. A striking similarity lies, then, between Hahnemann and Ritter in their self-experimentation with extremes.[74] Hahnemann's own descriptions of the effects of Peruvian bark, for instance, are unintentionally harrowing: he writes how first his fingertips became cold, he became limp and sleepy, then his heart began to beat, his pulse became hard and fast; followed by unbearable fearfulness, trembling (though without dread), a weakness in all the limbs; then a throbbing in the head, red cheeks, thirst, etc. (see Bayr 25–6). He recognized that these were all symptoms of malaria.

Goethe's characterization of Ritter as a pure empiricist, of course, is highly paradoxical: for the author of *Physik als Kunst*, empirically grounded experience served as proof of higher forces beyond the realm of the senses.[75] Novalis correctly assessed: "Ritter certainly seeks to reveal the true universal soul of nature. He aims to learn to read the visible and ponderable lettering, and to explain the state of the higher mental powers" (2: 816, #368). Through his experiments, Ritter believed he had discovered the galvanic principle enlivening all of life, from the worm to the human: "Each body part, as simple as it be, is to be regarded as a system of many infinitesimal galvanic chains . . . But is it different in the human body than in the skin of the worm? . . . The foundational proof is the continuous activity in the continuously linked chains" (*Beweis* 158). The invisible realm that for Goethe possessed "neither form nor limit" was for Ritter foundational proof ("Das Begründete"). As we shall see in the next chapter, Hahnemann's empirical observations led him to postulate that the exponentially diluted homeopathic remedy possesses an immeasurable spiritual, dynamic power, similar to Ritter's galvanic "continuous activity." In other words, like Ritter, he creates, with true Romantic faith and exuberance, what he claims only to observe.

"The Genius of Self-Poisoning"

In closing, another paradox in homeopathy still begs to be resolved: despite, as seen in this chapter, Hahnemann's belief in the totality of the individual, that is, how he investigates and treats the psychophysiological nexus in its entirety, the result is a self as a fragmentary totality. Instead of an explanatory diagnosis of the disease, Hahnemann replicates a welter of symptoms in his handwritten patient journals, protocol books, and repertories, as well as in the published *materia medica*. For instance, Kathrin Schreiber counts 408 symptoms listed for *bryonia*, 574 for *ignatius*, and 908 for *nux vomica* in the *Reine Arzneimittellehre* alone (29). Posed as a question, the issue is this: how does one respond to these bewildering and eclectic assemblages of symptoms? As opposed to the trimness and organization of the *Organon*, the form of these other works is open-ended and unlimited. How can we understand this discrepancy in terms of discursive practices around 1800?

In her study of the female patients who consulted with a doctor in Eisenach in the eighteenth century, with refreshing honesty about the limitations of the scholar, Barbara Duden openly admits that she feels like an intruder looking into the invisible corporeality of the patients (202). How does one begin to make sense of their descriptions of their agonies and afflictions?[76] One approach to this dilemma of the medical historian is to accentuate the Foucauldian disciplining and self-monitoring of the body in the attempt of physician and patient to control the waywardness of the body.[77] Even Hahnemann's voluminous collections, including above all the *Krankenjournale*, appear to want to render corporeality "legible" via his own script; he mediatizes the body through his various compendia. I would like to propose a different model, however, one that would accentuate the illegibility of corporeal pain that is reproduced in the interminable listing of symptoms and that strikingly contrasts with the purported efficacy of the spiritually enlivened remedy, the simplicity of the dietetic regimen, and the elegance of the principle *similia similibus curentur*.

The written results of Hahnemann's monitoring seem anything but disciplined and systematized (apart from the ordering of symptoms from head to toe, preceded by notation of the mental condition). They seem rather to obey Walter Benjamin's characterization of the "fanaticism of the collection" (1.1: 364). Hahnemann seems to indulge in an anti-reading because he does not study beneath the skin. Because they have no referent to nosological classification, his micro-perceptions display a hallucinatory quality. They are also tremendously energetic. He assembles an illegible, prodigious amount of information about infinitesimal variables. In the previous chapter, I compared this concept of the body as infinitely producing symptoms, unorganized into circulatory, digestive, or nervous systems, to Deleuze's Body without Organs. This somatic image comes into conflict, though, with the paradigm discussed in this chapter of integrated selfhood. What does such incongruity entail for Hahnemann's system? First, in marked contradistinction to this aggregation of physical and mental woes, and as if to contain them, the recommended care of self is ascetic and minimalist. Second, in outlining a fail-proof cure, Hahnemann controls contingency with a totalizing system of well-designed principles. The theory of *similia similibus curentur* is pristine, self-contained, and rigorously coherent. He coped with the unpredictability of the course of an illness with faith in a dependable healing method.

Put differently, given his view of the radical miscellany of disease, Hahnemann had to devise a counter-hegemonic explanatory system of illness and cure. And what better way to introduce rigour than to theorize the very inconsistency of facts as well as the fact of irreconcilability. In typically witty fashion, Friedrich Schlegel wrote in the *Athenäums-Fragmente* of 1786: "For the mind it is equally deadly to have a system and to not have a system. It will probably need to decide to bring both together" (*Kritische Ausgabe* 2: 173, #53). Two years before, Novalis similarly advocated that philosophy, to be free and unending, convert asystematicity into a system (2: 200, #648). And, indeed, the Jena Romantics devoted themselves to articulating a consistent theory of asystematicity. Their literary project moved away from a poetics based on strict genres to one where each text was unique and demanded individual interpretation. There was no standardized, generic way of reading. The genres that come to the fore in Romanticism are thus non-prescriptive. One could even call them non-genres – the fragment, the essay, and the mixed genre of the novel with its embedded fairy tales and digressions. Novalis and Friedrich Schlegel, in particular, theorized extensively about the Romantic fragment. It was to allude (*hindeuten*) without offering up definitive interpretation (*Deutung*) and revelled in surface, extraneous, or marginal observations. It hid more than it revealed and intentionally lacked context and cohesiveness.[78] Moreover, each fragment was, like an individual, complete unto itself, as Schlegel famously put it, like a hedgehog curled up into itself and separated from the surrounding world (*Kritische Ausgabe* 2: 197, #206). Returning to homeopathy, then, in its *theory*, every individual is unique. In *praxis*, this anti-system worked to let Hahnemann record the symptoms of each patient as distinct, inimitable, and utterly fragmentary. Hahnemann then systematically collected these fragments, as much as Novalis voluminously did, for instance, in *Das Allgemeine Brouillon* (*The Universal Notebook*) of 1798–9. The goal of Novalis's unfinished Romantic encyclopedic project was, in fact, a strikingly similar one: the juxtaposition of fragmentary knowledge from different sciences and sources was to allow for the discerning of analogies and connections.

Another major implication of the extensive registering and recording of the fragmentary, micro-perceptions of the body, as seen both in Hahnemann's *materia medica* and his *Krankenjournale*, is that the anamnesis takes precedence over medical diagnosis, in fact, it overrides its necessity entirely. If Hahnemann can stress solely the listening to his patients' complaints, he is absolved of the responsibility of a diagnosis.

He indeed creates a theory of why identifying a disease does not matter, namely, that each individual is unique in his or her manifestation of illness. Therapeutic recommendations to follow a strict diet and to self-monitor, especially as seen in patient correspondence, also function as a substitute for diagnosis. The case of Princess Luise, who extended her treatment over a considerable length of time, even suggests that homeopathic therapy is an elite leisure activity. Its increased popularity and commercialization over Hahnemann's lifetime, as was particularly the case during his Paris years, where such famous individuals as Paganini and Lord Elgin sought his counsel, would indicate that accurate medical diagnosis would matter less and less and the patient's recounting of the body's variable afflictions more and more. Annette von Droste-Hülshoff reported in 1830 that, having lost several moneyed patients to Bönninghausen, local physicians sneeringly said Hahnemann was a doctor for refined folk, especially women who liked to chat about literature and fine art (106). Whether or not she was being self-ironic is unclear.

Finally, there is another way by which Hahnemann takes advantage of the hyper-signification of the individual body. His self-provings, which initially resemble acts of reckless self-poisoning, are actually a domestication of toxicity and a means to manage and regulate his patients' incoherent ailments. By simulating through drugs their illnesses, Hahnemann parses them in advance, in place of their bodies, and in preparation for the consultation. He serves as a medium for his patients and positions himself to empathize with their suffering. Bettina Brockmeyer has perceptively written about the self-performance through letter writing of Hahnemann's patients, above all, in their address to the physician. One can equally speak, though, of Hahnemann's anticipatory performance of the other. By strategically performing in advance what his patients experienced, he positioned himself as their "hyper-simile" (Sloterdijk 33). At the same time, he grants the individual power by encouraging him or her to monitor and narrate corporeal vulnerability, the patient's symptoms ultimately ratifying Hahnemann's own experience and his system, for the selection of the remedy is based on matching these sufferings to ones that he has previously undergone. *Their* experiences are instrumentalized to confirm *his* prior tabling of symptoms in the *Reine Arzneimittellehre*. In this fashion Hahnemann can contain the chaotic and infinitely variable plethora of baleful symptoms, regulate the wayward body, and at the same time confirm his position as an exceptional medical authority.[79]

The cult of genius is not unique to medicine, of course. In *When Physics Became King* Iwan Rhys Morus writes about the early nineteenth century: "Natural philosophy required genius. Only a genius – an inspired individual with access to unique reserves of imagination and intuition – could peer beneath the fractured surface of appearances at the transcendental reality beneath . . . In the arts, literature, and music, as well as natural philosophy, being a genius was very much in vogue" (55). A child of his time, Hahnemann predictably lauds his own achievements in terms of the solitary mind of the Romantic scientist voyaging into the unknown. Still, he excelled in stylizing himself as such. He positioned himself as the inventor of the only medical cure to guarantee success; he claimed to "peer beneath the fractured surface of appearances" to divine the appropriate remedy; and he insisted that patients travel to him rather than vice versa. Above all, in Peter Sloterdijk's concise and virtuosic turn of phrase, Hahnemann was the "genius of self-poisoning" (33).[80]

Chapter Three
THE LAW OF MINIMUM

"Every grain / Is sentient and both in unity and part"

As discussed in chapter 1, in his inaugural essay of 1796, "Versuch über ein neues Prinzip zur Auffindung der Heilkräfte der Arzneisubstanzen," where he introduced the notion of *similia similibus*, Samuel Hahnemann attacked the horse medicine of his day – the prevalent prescribing of significant bloodletting, copious amounts of opium, and drastic emetics and purgatives. By contrast, he noted the effectiveness not just of moderation, but specifically of the small dose. By 1802 he announced his principle of successive dilutions, or what was to be termed homeopathy's Law of Minimum. Respecting it, over time he gradually decreased the ratio of concentration of the mother tincture, which paradoxically meant that the remedy would be stronger acting. The impact of the catalyst was present even though the toxicity of the substance had disappeared, precisely because of the dynamic intensity that it developed.

According to this principle, the homeopathic pharmacological remedy is dynamized or potentized by a series of dilutions on the scale of one part of original substance to nine parts of alcohol (or water) or one part to ninety-nine. This computation creates either the decimal scale ("X") or the centesimal scale ("C"). Potencies are named according to the number of times they have been diluted in the X or C scale (e.g., 6X, 30C, 100C). Less of the original substance means a more profound effect as a remedy becomes increasingly more energetic: the higher the number of dilutions the deeper the remedy acts. The living spirit within it becomes ever more operational. Although it was not published until 1921 by the Hahnemann biographer Richard Haehl (1873–1932), the sixth edition of 1842 of the *Organon* discusses even the 50 millesimal potency, which also goes by the name "Q-potency" (from Lat. *Quinquaginta milia* = 50,000). In §270 of this edition Hahnemann prescribes how to dilute a substance infinitesimally, at which point it will resolve totally into its individual, spiritual essence.[1] Chapter 3 is devoted to retracing both this development in Hahnemann's thought and the discursive framework that contributed to his concept of dynamization.

Although with the Law of Minimum Hahnemann established a new medical theory, his innovation cannot be contemplated outside related systems of thought current at the time. Take, for example, the following passage from *Queen Mab* (1813) by Hahnemann's contemporary Percy Bysshe Shelley (1792–1822):

> Throughout this varied and eternal world
> Soul is the only element: the block
> That for unaccounted ages has remained
> The moveless pillar of a mountain's weight
> Is active, living spirit. Every grain
> Is sentient and both in unity and part,
> And the minutest atom comprehends
> A world of loves and hatred. (776)

Shelley demonstrates a belief in an "active, living spirit" or "Soul" that permeates even the inorganic world, turning either a mountain or the minutest atom into sentient being: the part is just as enlivened as the whole of nature. Hahnemann, too, takes recourse to the notion pervasive at this time, in Shelley's words, of an "active, living spirit." This energetic force is present not only in the human body but throughout nature, indeed even and especially at the infinitesimal molecular level of the remedy or, as Shelley puts it, in the "minutest atom."

Already in "Heilkunde der Erfahrung" of 1805 Hahnemann writes: "The *dynamic* action of medicines, like the vitality itself, by means of which it is reflected upon the organism, is almost *spiritual* in its nature" ("The Medicine of Experience," *Lesser Writings* 466). What is unique about the invention of homeopathy is that in conjoining the concept of a vitalist force in nature with the idea of infinitesimal dosage, homeopathy develops an organicist view of nature, prevalent at the time of Romanticism, into a minimalist one. The substance expresses itself wholly in the minuscule part, and the human body responds fully to the ever so slight prompting of the treatment. The plethora of symptoms is remediated by the simplicity of the infinitesimal. Moreover, according to the later editions of the *Organon* and in the prefaces to the later volumes of the *Reine Arnzeimittellehre*, Hahnemann postulated that a substance would be not just still present but in fact activated after exponential dilution, as well as by trituration (grinding an insoluble dry ingredient with milk sugar) and succussion (vigorous shaking of a liquid). "This development of the spiritual power of medicines to such a height by means of the multiplied and continued *trituration* and *succussion* of a small portion of medicinal substance . . . deserves incontestably to be reckoned *among the greatest discoveries* of this age" ("How Can Small Doses Still Possess Great Power?" *Lesser Writings* 729–30 and *Materia Medica Pura* 44). And just as the vibrational energy within the remedies could be potentiated, so too could it be malformed: by the 1833 fifth edition of the *Organon* Hahnemann militated against transporting homeopathic remedies for fear that their powers would be altered by unwanted shaking.

Significantly, Hahnemann says that it is not a question of an "equal portion of the drop of medicine" contained in every part of the mixture and distributed "uniformly *through the whole mass*" (*Materia Medica Pura* 44). He speaks instead of the "development and liberation of the dynamic powers of the medicinal substance" (43), of "developing and intensifying the dynamic forces of medicine" (44). "Medicinal substances," he writes, "are not dead masses in the ordinary sense of the term, on the contrary, their true essential nature is only dynamically spiritual – is a pure force, which may be increased in potency almost to an infinite degree" (46). In other words, as he already pointed out in 1805, the action of the medicine on the body is not chemical but *"purely dynamic"* ("The Medicine of Experience," *Lesser Writings* 465).[2] What one sees here is a reflection of Hahnemann's contention that the early-nineteenth-century isolation of various plant alkaloids, which gave rise to pharmacology, has no applicability to homeopathy. He specifically states that "[the] water or oil, distilled from the plant, or the resin obtained from it, is certainly not its active principle" ("Examination of the Sources of the Common Materia Medica," *Lesser Writings* 675). Instead, he posits the "dematerialization" that occurs in this *"almost illimitable"* development of the powers of medicinal substances ("Remarks on the Extreme Attenuation of Homoeopathic Remedies," *Lesser Writings* 763). The process of dilution involves not so much the discrete still maintaining a property of the whole, but the particular intensifying the whole. As Harald Walach has pointed out in his essay "Magic of Signs," Hahnemann himself does not speak in terms of an informational or content transfer in the remedy. Another way of putting it is there is no molecular memory effect or causal, material link between the original plant substance and the dilution. Instead, the action of the medicine is, in Hahnemann's own words, spirit-like or dynamic.

This conviction in the catalytic power of the infinitesimal dilutions sees itself grounded in material reality, however contradictory this may seem. The pure current running through the infinitesimal dose is testimony to its vibratory *life*: it is concrete and materialist in its corporeal effectiveness. The activity in the dose becomes a creative agent that animates, directs, and rearranges bodily health and impels restitution. The force in the remedy flows across material boundaries and thus cannot be conceived as inert matter. Furthermore, the impulse in the microdose is only evident in interaction or communication with the life force in the human being. In fact, its intensity only seems to emerge through interaction with the body, whose own vital life force (*Lebenskraft*) responds and is in turn enhanced. Another way of stating this enigma is that activity and spirituality can be seen as operative par excellence within the mineral or botanical medicinal substance. These minute, varied substances represent the infinite and infinitely active manifold of the natural world. The body, too, is part of this vitalistic flow. Because of the realism inherent within homeopathy, one cannot assert that *Lebenskraft* is identical with the soul: whereas the soul can be said to exist without a body and represents opposition to the body, *Lebenskraft* has existence within the body. This dynamic power or intensity is life itself.

What is also curiously paradoxical is that, precisely at the moment the ingredient in a remedy becomes most immeasurable, homeopathy stakes its claim in close observation, empirical testing, and verifiable evidence. Hahnemann writes in the *Organon*: "By a mere effort of the mind we could never discover this innate and hidden *faculty* of medicines – this spiritual *virtue* by which they can modify the state of the human body and even cure disease. It is by experience only, and observation . . . that we can either discover or form to ourselves any clear conception of it" (*Organon* 103, §20). Typical for Romantic science (about which I shall shortly say more), homeopathy asserts through experimentation to demonstrate something invisible. Hahnemann claims to bear tangible witness to what was not directly manifest. What he fails to acknowledge is that there could be tension between empirically based scientific experimentation and the philosophy of vitalism. Instead, to justify this leap of faith, Hahnemann rhetorically invokes Goethe as an authority and cites his derision of unbelievers from *Faust, Part Two*:

> Daran erkenn' ich die gelehrten Herrn!
> Was ihr nicht tastet, steht euch Meilen fern;
> Was ihr nicht fast, das fehlt euch ganz und gar;
> Was ihr nicht rechnet, glaubt ihr, sei nicht wahr'
> Wahr ihr nicht wägt, hat für euch kein Gewicht;
> Was ihr nicht münzt, das, meint ihr, gelte nicht. (qtd. in "Remarks on the Extreme Attenuation of Homoeopathic Remedies," *Lesser Writings* 764)

> By this I recognise a most learned lord!
> What you can't feel lies miles abroad,
> What you can't grasp, you think, is done with too.
> What you don't count on can't be true,
> What you can't weigh won't weigh, of old,
> What you don't coin: that can't be gold.[3]

This rhetorical sleight of hand shows Hahnemann on the defensive. And, to be sure, the Law of Minimum is the most contentious issue in homeopathy. I cannot resolve here whether the efficacy of the minimal dose is demonstrable fact or a matter of faith. But what I do want to stress is that Hahnemann's conceptualization and framing of this law need to be seen in terms of discourses current at the time. In this chapter I will thus rehearse the successive stages Hahnemann went through to arrive at the infinitesimal dose as well as the amount and frequency of its dispensing. I shall also place the Law of Minimum on a continuum with other dietetic, minimalist care practices, as well as with vitalistic notions in the life sciences around 1800. I shall look at the comparisons Hahnemann initiates between, on the one hand, homeopathy and mesmerism – arguably the most sensational medical practice of the era – and, on the other hand, homeopathy and infinitesimal mathematics. I also want to raise the question of how the small dose differs conceptually from vaccination, a practice gaining more and more recognition around 1800 as a preventative against smallpox. Finally, I shall examine parallels between the terminology of potentization that Hahnemann and the German Romantics share.

From the Small to the Infinitesimal Dose: Hahnemann's Development of Thought

In his 1796 essay, Samuel Hahnemann railed against the palliative use of drugs for their merely temporary effect, which led to higher doses and unwanted side effects. He writes, "In chronic diseases it only gives relief at first; subsequently, stronger doses of such remedies become necessary, which cannot remove the primary disease, and thus they do more harm the longer they are employed" ("The Curative Powers of Drugs," *Lesser Writings* 262). He offers a strikingly accurate assessment of mercury, a widely used agent against syphilis even until 1943, when penicillin was first found to be an effective treatment: "It is often very difficult for the practitioner to distinguish the chronic mercurial disease from the symptoms of syphilis; and thus he will be asked to consider symptoms as belonging to that disorder, while they are only mercurial, and go on treating them with mercury, whereby so many patients are destroyed" (ibid. 286).[4] The more he devoted himself to studying chronic disease, the more Hahnemann was convinced that the overprescription of medication was the chief cause of illness, especially in the long term. Already in 1813, he wrote of the first "unmistakable laws of nature": "*the living organism is incomparably less capable of being affected by natural diseases, than by medicines*" ("Spirit of the Homoeopathic Doctrine of Medicine," *Lesser Writings* 627). Then, in 1831, he refers to how patients have been "injured to the verge of incurability by the allopathic exterminating art" ("Allopathy: A Word of Warning to All Sick Persons," *Lesser Writings* 751): chronic diseases are produced either by "the long-continued use of large doses of strong medicines unsuitable for the disease" or by a "simple medicament employed for a length of the time in frequent, large doses" (ibid. 750).

Initially, Hahnemann's goal was to find a balance between minimal toxicity and maximal benefit, as well as, as we saw in the last chapter, to match the medicine to the individual patient's needs. In both respects, he was reacting against Brunonian medicine, designed to restore the weakened body to health by administering stimulants. Hahnemann was not the only thinker at the time to attack over-medication, consequently to recommend minimalizing dosages, and to do so within the wake of Haller's notion of irritability (*Reizbarkeit*). Hufeland, for example, felt that overstimulation (*Ueberreitzung*) was characteristic of his age ("Geschichte der Gesundheit"). If the modern person was exposed to too many stimuli, then the physician needed to let the body restore its own equilibrium. In his medical notations, Novalis likewise expressed the opinion that sensitive persons were already exposed to too many stimuli and as a result must be prescribed very little – indeed extremely diluted – medication (2: 595, #536). In this instance, Novalis is speaking about narcotic remedies, opium being regarded as a stimulant.[5] At the same time, Novalis also made fun of Hufeland's macrobiotics for advocating "the most diluted life" (2: 557, #437). Instead, he recommended *"The art of living – against macrobiotics"* (2: 415, #445). This "art of living" demanded the concerted dedication of the individual to fashioning his own life (2: 388, #343).[6] A Brunonian adherent, the physician Karl Friedrich Burdach also challenged that Hufeland's approach was too passive, whereas he encouraged a "polybiotics" or active existence through increased pleasure.[7] Upholding minimalist care, Hahnemann would have sided with Hufeland in these Brunonian-inflected debates over the increased or decreased need for stimulus, but he did not subscribe to Hufeland's therapeutic nihilism and insisted upon the wise intervention of the homeopath. Still, homeopathy can be seen as a type of provocation or *Reiz* that gently induces the organism's vital life force to recalibrate.

Novalis offers some of the most striking because counter-intuitive passages on the workings of drugs: "Medicine is indeed the art to kill . . . All drugs are, due to their effectiveness, harmful" (2: 500–1, #143). He also opined that several constitutions could not tolerate medicinal and food substances in concentrated form, even when administered in droplets (2: 542, #374). As this passage demonstrates, like Hahnemann but in contradistinction to Brown, Novalis championed individual patient care. He also, like Hahnemann, recognizes the peculiar qualities of the pharmakon: it can be both toxic but also curative once diluted. Novalis continues: "Dilution is necessary here, and then, properly diluted, a much greater quantity can be used without causing damage. Therein, I believe, lies the peculiar characteristic of poison" (2: 542, #374).[8]

Goethe arrived at the strikingly similar, paradoxical conclusion about the pharmakon. In the same year that Hahnemann came out with his inaugural essay, Goethe reports a conversation with Christoph Martin Wieland (1733–1813), in which the latter complained that young people were imbibing too much tea, exposing themselves to its debilitating properties. Goethe responded with an argument about the relativity of toxicity: tea does not solely weaken but can either strengthen or weaken. He then exclaimed: "There is no poison! . . . It all depends on the dosage. Even champagne can turn into poison!" (GA 22: 251). Around this time, in the year 1795, the physician Carl Christian Heinrich Marc (1771–1840) wrote in almost exactly the same words, "There is no poison, i.e., no absolute poison" (31). He points, for instance, to the well-known fact that the same toxin can have varying effects depending on the type of animal species exposed to it (30). Marc turns this argument about relativity on its head, though, and says that it is not only increased medicinal doses that create toxicity; many curative substances are also poisonous insofar as they have noticeable effects in very small doses (236). Even the most powerful poisons can through artificial modification become more benign and demonstrate healing powers (241). Thus, if Hahnemann was not the only physician to criticize the toxicity of medicines intended to cure, he was also, conversely, not the only one to see in the toxic substance a potential remedy.[9]

Interestingly, Marc interchangeably uses the terms poison (*Gift*), medicine (*Heilmittel*), and illness (*Krankheitsgifte*), by which he means that diseases, such as contagious ones, have a kind of poisonous, aggravating effect on the body – they function as *Reiz*. Hahnemann, too, in his important essay of 1805, "Heilkunde der Erfahrung," wrote: "Every disease is owing to some abnormal *irritation* [*Reiz*], . . . which deranges the functions and well-being of our organs" (447). But whereas for Marc illness functions like an unnatural toxin, for Hahnemann the toxin literally functions like the illness: it simulates it. In other words, the minimal homeopathic dose of a toxin produces in the body a simulacrum of the original morbid symptoms, causing the body to fight these counterfeited symptoms and thereby overcome the entire disease. He therefore writes a few pages later: "*It is only by this property of producing in the healthy body a series of specific morbid symptoms, that medicines can cure diseases, that is to say, remove and extinguish the morbid irritation* [*den Krankheitsreiz*] *by a suitable counter-irritation* [*Gegenreiz*]" ("The Medicine of Experience," *Lesser Writings* 451).[10]

Here again Hahnemann is not alone. Novalis, too, formulated the hypothesis that one illness can cure another: a localized malady often is cured via the stimulation (*Erregung*) caused via a general illness, and vice versa (2: 706, #1057). He even whimsically toys with the idea that one can cure pain through tickling (2: 708, #1071). Time and again via his tendency to think in polarities, or what John Neubauer called his bifocal vision, Novalis speaks in his fragments of how illness and death can lead to life, as when he writes that death is a means to life (2: 350, #166) or in an entry on medicine about the uses of each disease – and the poetry thereof (2: 475, #12). The curative function of illness leads him even to conceptualize a *"Poëtik des Übels"* (2: 628), which David Farrell Krell translates as the "Poetics of the Baneful." In the same lengthy fragment he rhetorically asks the counter-intuitive questions of whether disease could be the means to a higher synthesis and whether everywhere the best things start with disease. He even postulated that hypochondria paved the way to somatic self-knowledge, self-control, and self-enlivening (2:397, #387). "Absolute hypochondria – hypochondria must become an *art* – or education" (2: 403, #420). Illness thus holds the potential for greater self-awareness. In sum, although Novalis stops short of developing an entire medical system, as does the founder of homeopathy, he does postulate the remedial powers of illness, even curing, in Hahnemann's words of 1796, "a chronic disease by superadding another" ("The Curative Powers of Drugs," *Lesser Writings* 265).[11]

Meanwhile, though, Hahnemann's thinking about the healing properties of minimally administered toxins continued to evolve. Although he at first thought that the lesser the quantity of a medicine, the more its potency would be diminished (von Hörsten 73), over time the founder of homeopathy reversed the relation. To rehearse Hahnemann's move from the minimization of toxicity to the dynamization of effect, I rely on the factual compendium of Stefan Mayr's *Herstellung homöopathischer Arzneimittel* and Matthias Wischner's *Fortschritt oder Sackgasse?* (2nd edition, 2006) on Hahnemann's late writings from 1824 to 1842.[12] Mayr argues that it is difficult to retrace in Hahnemann's case journals how and when he arrived at this shift, for after 1804 he mentions less and less the dilutions he administers (20). By 1805 such notations disappear (20). The case journals of 1813–14 mention briefly a dilution of 1:1000, but not again (25). Varady surmises that this absence is the result of Hahnemann having arrived at a standardized dose. Mayr, by contrast, suggests that it was in order to avoid potential criticism of such a speculative procedure. Varady does note, however, that in 1801, in an essay published on scarlet fever, Hahnemann gave his first detailed instructions about the amount, frequency, and duration of administration for *belladonna*. He also mentions that he used 1/24 of 1 million of a grain for each drop, an admission that shocked the medical community and led to attacks. According to Varady, this hostile reception led Hahnemann not to take up the discussion of the degree of dilutions and their prescription openly until 1809. Mayr's conclusion could thus be supported by her observation.

It wasn't until 1830 that Hahnemann defined the dosage of C30 as the standard potency (*Regelpotenz*), as seen in the §§270–1 of the 1833 fifth edition of the *Organon*, a dosage that today is still regarded as the standard potency. In §283, however, he does mention the occasional necessity of higher potencies. As to sniffing a remedy, Hahnemann first recommends it for exceptional cases in the *Reine Arzneimittellehre* of 1818. Then, in §288 of the fifth edition of the *Organon*, aromatic exposure becomes standardized procedure (Wischner, *Fortschritt oder Sackgasse?* 244–7). But even as early as 1801, in an essay published in Hufeland's *Journal*, entitled "Ueber die Kraft kleiner Gaben der Arzneien überhaupt und der Belladonna insbesondre" ("On the Power of Small Doses of Medicine in General, and Of Belladonna in Particular"), Hahnemann gives the wonderful example of smelling soup. His point is to illustrate how much more sensitive the sick person is to the powers of medicines: "What an enormous quantity of freshly made soup it would take to excite a healthy stomach to violent vomiting! But look, the patient ill of an acute fever does not require a drop for this purpose; the mere smell of it, perhaps the millionth part of the drop, coming in contact with the mucous membrane of the nose, suffices to produce this result" ("On the Power of Small Doses of Medicine," *Lesser Writings* 388). Then, at the end of his life, instead of alcoholic dilution or sniffing, Hahnemann recommended putting the potentiated dilution onto a little globe of sugar of a definitive size, which the patient was to dissolve in water, developing its potency even more (see the preface to the second edition of the third volume of *Die chronischen Krankheiten* of 1837). It goes without saying that Hahnemann stipulated that the homeopath prepare his own remedies in order to guarantee their purity (*Organon* 214, §264).[13] On this topic he grew more and more adamant with age.

The case journals also give ample testimony to how frequently Hahnemann substituted placebos. Schuricht postulates that Hahnemann wanted to give himself time to find the appropriate remedy by the next visit (*D16* 55). Varady surmises that, in addition to waiting for the effect of allopathic medicines to wear off, Hahnemann used placebos to test the accuracy of what patients were saying about the results of a dosage. He also did not want to cater to patients who demanded drugs. She points out that between 1803 and 1805 he prescribed placebos 25 per cent of the time (*D5* 44–68).

As to the frequency of dosage, the above-mentioned article of 1801 on *belladonna* and scarlet fever give some indication of how Hahnemann might have determined it. He notes that the "peculiar action of this plant does not last above three days" ("Cure and Prevention of Scarlet-fever,"

Lesser Writings 380). Consequently he repeats the dose every 72 hours. He also, incidentally, prohibits the eating of sour fruits and vinegar, because they significantly increase the action of *belladonna* (ibid. 382), an effect he wanted to avoid. Then, in 1805, Hahnemann instructed: "The remedy must be given in smaller and smaller doses, repeated at longer intervals, to prevent the occurrence of a relapse; if the first, or the first few doses have not already sufficed to effect a cure" ("The Medicine of Experience," *Lesser Writings* 454).[14] After 1813 Hahnemann increasingly dispensed only one remedy at a time and only switched if necessary: the single dose should be given as much time as possible to display its effectiveness (Schreiber 32). Indeed, up until the fourth edition of the *Organon* (203, §242) Hahnemann stated that in both acute and chronic diseases only one dose is to be given until its effectiveness runs out. But then in the fifth edition he takes backs this recommendation, which he now regards as too dogmatic: §§246–7 specify that the medicine should be taken in varying intervals according to the needs of the patient, even every five minutes. By 1837 he is changing potencies slightly on a daily basis. Consequently, in *Organon 6* he establishes as a rule the daily meting out of medication (Wischner 345). The first two editions of *Die chronischen Krankheiten* stipulate taking the remedies in the morning, but the preface to the second edition of the third volume of 1837 recommends the evening before going to bed, because there is less chance then of external interference (*Die chronischen Krankheiten* 3: xiii).

As to which part of the body the remedy needed to be administered to, Hahnemann wrote:

> The contact of the medicinal substance with the living, sensitive fiber is almost the only condition for its action. This dynamic property is so pervading, that it is quite immaterial what sensitive part of the body is touched by the medicine in order to develop its whole action, provided the part be but destitute of the coarser epidermis – immaterial whether the dissolved medicine enter the stomach or merely remain in the mouth, or be applied to a wound or other part deprived of skin." ("The Medicine of Experience," *Lesser Writings* 466)

Lebenskraft

After this basic overview of Hahnemann's application of the Law of Minimum, it is crucial to devote time in this chapter to discussing how he envisaged dynamism – how the body and the medicine interact holistically and organically through shared vital energies. In assuming the presence of both *Lebenskraft* within each person and the spirit-like effect of the remedy, Hahnemann was operating within the realm of the invisible. Yet, through homeopathically prompted auto-healing he claimed to be offering material evidence for incorporeal, even ethereal forces. Although it was a mere postulate, *Lebenskraft* represented utter naturalness – the body's ability to heal without harsh, artificial intervention. And, however undetectable and imperceptible the dynamic essence in the remedy might be, it marked the sign of organic vitality and the transmissible intelligibility of plant life. Thus, although, paradoxically, these intensities and powers were unverifiable and imponderable, they underpinned empirical reality, in fact allowed Hahnemann to claim scientific truth and accuracy in his testing.

Linguistically speaking, the very word *Lebenskraft* was attractive precisely because it gave voice to both a secularized notion of independent reality or *life* and a benevolent yet powerful autonomous *force* operative within nature. It was general enough to be claimed to subtend everything in the organic and inorganic world, yet specific enough to be said to self-regulate all the internal corporeal processes. Dynamic processes were both internal and external, everywhere and nowhere. Most importantly, Romantic vitalism presented nature as a unified whole that could not be broken down into constituent parts. As Hahnemann put it, the "organic union" and "full development of life . . . (which can only be defined by the term *vitality*) . . . cannot be judged or explained by any other rule than that which itself supplies; therefore, by none of the known laws of mechanics, statics or chemistry" ("Observations on the Scarlet-fever," *Lesser Writings* 489).

The concept of *Lebenskraft* is a crucial one for Hahnemann because, although he does not begin to apply it with frequency until 1824 (in the later editions of the *Organon*, substituting it for *Natur* or *Leben*), it grounds the presuppositions and principles of homeopathy. Even if Hahnemann does not exercise the term *Lebenskraft* in his earlier writings with the same frequency he does later in life, still, from the start he propounded the unity and totality of the organism and its ability to react as a whole to a substance. He therefore, as we saw in the previous two chapters, lists every symptom of an illness as being reflective of the whole organism, which responds, as an entirety, to the remedy. His symptom-based semiotic theory and his recourse to *Lebenskraft* go hand-in-hand in another way. Hahnemann requires, conceptually speaking, a unified driving, regulatory force or principle to overcome the dispersion and dissolution of these manifold symptoms. The unaccountable dark interior of the body (inexplicable, as homeopathy refuses to ascribe to a pathology or a nosology) leads him not only to observe and enumerate external, visible indices but also to posit an invisible yet all-pervasive *Lebenskraft* that, when stimulated, could overcome the incomprehensible illness. *Lebenskraft* alone, once prompted by the homeopathic remedy, was enough to counter the fall into sickness, degeneration, and decay catalogued so meticulously in his protocol books, case journals, and *materia medica*. Its activity is needed to prevent all *their* ever-growing compendia from getting out of hand. In other words, precisely because of the sheer, overwhelming reality of disease symptoms, *Lebenskraft* is conceivably introduced more and more into Hahnemann's writings as a potential guarantor of inner equilibrium. As a unifying principle, *Lebenskraft* deals with the ever-widening chasm of disparate disease phenomena.

To understand these paradoxes more fully – and to see them as representative of a dilemma at the core of Romanticism – I shall elaborate not only on the medical tradition that fed into Hahnemann's use of *Lebenskraft* – and here I shall reserve this particular term to how Hahnemann deployed it, namely, in reference to the human body – but also on how general dynamic principles played out in wider contemporaneous debates in the life sciences. Hahnemann evoked a belief in unseen vital powers very familiar to his readers. Vitalism was debated among physiologists, naturalists, philosophers, and physicists alike. More specifically, I shall examine how, like the Romantics, Hahnemann moves outside the epistemological limits that Kant set when scrutinizing the regulatory principle of unseen but vital forces operative in nature. Hahnemann's deployment of the concept of *Lebenskraft* also explains why he considered mesmerism to be a form of homeopathy.

The term *Lebenskraft* was coined in 1774 by the physician and botanist Friedrich Casimir Medicus (1736–1808) and gained widespread use in the eighteenth and early-nineteenth centuries. One of the most influential thinkers of the period, Johann Gottfried Herder, extolled the powers (*Kräfte*) that underlay human and animal life. "Physiology," he wrote, "of the human or any animal body [is] nothing but a *realm of living forces* . . . Everything that we call matter is therefore more or less in itself enlivened; it is a realm of active forces that according to their nature and their relationships form a complete whole [ein Ganzes bilden], and not only appear as such to our senses . . . One force dominates: otherwise everything wouldn't be unified, not a whole [kein Eins, kein Ganzes]"

("Gott, einige Gespräche" 4: 774). German scientists, including Johann Friedrich Blumenbach (1752–1840), Gottfried Reinhold Treviranus (1776–1836), Carl Friedrich von Kielmeyer (1765–1844), and Johann Wilhelm Ritter – demonstrated a firm belief in a vital, organicist, unifying life force. But this force raised more questions than it answered: Was vitalism to be explained as irritability (demonstrated in the muscles) or as sensibility (demonstrated in the nerves), as medical specialists after Haller were to investigate?[15] Blumenbach and Treviranus investigated infusoria and zoophytes under the microscope in order to trace what was invisible to the naked eye and yet underlay processes of generation, Blumenbach being the one to coin the term *Bildungstrieb* or *nisus formativus*. Not restricted to generative powers, vital life forces in the guise of electrical impulses were studied by Alexander von Humboldt (1769–1859) and Ritter in order to determine whether they were chemical or magnetic in nature. Did the electrical experiments conducted by Luigi Galvani (1737–98), Alessandro Volta (1745–1827), Humboldt, and Ritter produce animal energy, contact energy, or a combination of both? And how were these forms of energy related to human physiology?[16] Karl Rothschuh (*Physiologie* 187–90) points out that by 1801 the discovery of a galvanistic fluid was being linked to *Lebenskraft*, as reflected in writings by Johann Ferdinand Authenrieth (1772–1835), the founder of the medical clinic in Tübingen and infamous for the horrific masks with which he bound the faces of the mentally ill. Even as late as 1820, animal magnetism formed the basis of the physiology of Georg Prochaska (1749–1820). Hahnemann, as we shall see later, reinterprets mesmerism as a form of homeopathic practice.

In all these writers such dynamic processes were only a postulate which needed to be confirmed by observational inquiry. Because they could not be pinpointed, they invited scientists to hypothesize governing laws, as if these would then provide the key to nature. Although the belief in vitalism was premised on the hiddenness in nature, it invited experimental intervention so as to make unseen processes evident. Thus, on the one hand, vital activity represented nature embodied and hence embeddedness within the environment, yet, on the other hand, it was an abstraction from nature. As Joan Steigerwald has noted, scientists at the time interrogated the "border zones of life,"[17] between plants and animals, and between the lifeless and living. But if vital forces coursed through the entire world, then organic and inorganic nature shared the same substantiality. Vitalism expressed the dream of universality and the concatenation of being. Nature was seen as a whole, and one life force permeated all of her. As Herder wrote, "We do not have the senses to examine the innermost being of things; we stand on the outside and must observe. The more clear-sighted and still our gaze, the more the living harmony of nature reveals itself to us" ("Gott, einige Gespräche" 4: 778).

In the field of medicine around 1800, the physicians Christoph Wilhelm Hufeland, Andreas Röschlaub, Carl Arnold Wilmans (1772–1848), Conrad Joseph Kilian (1771–1811), Philipp Franz von Walther (1781–1849), and Johann Christian Reil were among those who subscribed to the notion of a life power that subtends optimal health. Xavier Bichat (1772–1802), who is known as the father of histology and the first to claim that tissues are distinct entities, wrote that "life is the embodiment of functions that resist death . . . permanent principle of reaction [Reactionsprincip] . . . This is the principle of life [Lebensprincip]" (1). The emphasis of these physicians was, of course, on health and disease, and thus shifted slightly from that of the biologists, such as Treviranus, Blumenbach, and Goethe, who examined and debated processes of generation and forms of natural self-organization. They also stand apart from the scientists Galvani, Humboldt, and Ritter, who inquired into the nature of animal energy. Much recent and prominent scholarship in the English-speaking world on Romanticism both in literature (Denise Gigante, Theresa Kelley, Robert Mitchell, Catherine Packham, and Sharon Ruston) and in the history of science (Timothy Lenoir, Peter Hanns Reill, Robert Richards, Joan Steigerwald) is devoted to this nonmedical aspect of vitalism.[18] The medical field, by contrast, remains surprisingly under-explored,[19] especially given that the notion of *Lebenskraft* was first and foremost developed in this arena.[20]

As deployed by physicians, the notion of *Lebenskraft* was indebted to the older sense that illness arose from an imbalance in the body. It goes back to the Neoplatonic belief that nature, although functioning according to laws inscrutable to mankind, was a vital whole that reflected the will of God. The human body was an organism that mirrored this unity and that had the natural capacity spontaneously and purposively to regain health. According to Hippocratic tenets, the task of the physician was to assist the *physis* in restoring its balance and maintaining well-being through minimal medical intervention and by advocating lifestyle moderation. As Roy Porter has summarized: "Classical medicine taught that the right frame of mind, composure, control of the passions, and suitable lifestyle, could surmount sickness – indeed, prevent it in the first place: healthy minds would promote healthy bodies" ("What

Is Disease?" 80). Abiding within this tradition, Hufeland writes: "Hippocrates, and all the physicians and philosophers of that period, knew no other method of accomplishing this end than by moderation; the use of free air and purer air; bathing; and, above all, by daily friction of the body and exercise" (*Art of Prolonging Life* 4).

Prior to these writers, the physicians who ascribed to a vital, indwelling spirit that regulated health included Paracelsus, who used the term *archeus*, which was later taken up by Jan Baptist van Helmont (1579–1644), and Georg Ernst Stahl, who preferred the term *anima sensitiva*.[21] These earlier doctors saw the corporeal wellspring of vitality as inseparable from the mysterious workings of the soul. Hence they were in opposition to a Descartian division of body from soul, a view inherited by Boerhaave, who spoke in iatromechanical terms. This dualism between the mechanistic and spiritual view of the body was overcome or superseded by another binary – Haller's division into irritability (via muscles) and sensibility (via nerves).[22] Inspired by Haller to bracket out the soul in his investigation of physiology, La Mettrie envisioned man as a machine. But Haller's new paradigm led more than anything to a framing of life in neither mechanistic nor metaphysical terms – the formulations of the *principe de vie* in 1772 by the Montpellier physician Paul-Joseph Barthez (1734–1806) and *Lebenskraft* in 1774 by Medicus. By the late eighteenth century the body was conceived to operate according to its own independent source of vitality, although in harmony with nature. Responding to this tradition in his 1808 essay "Ueber den Werth der speculative Arzneisysteme" ("On the Value of Speculative Systems in Medicine"), Hahnemann criticizes Stahl and van Helmont (ibid. 490) for their delusional fantasies and mysticism, but praises Haller and Blumenbach as being "the wisest among us" for their *Erfahrungsvitalitätskunde* ("empirical knowledge of vitality"; ibid. 493).[23] An organism's vitality is thus not attributable to a metaphysical power but is seen as a natural process.

The most famous medical proponent of *Lebenskraft* was Hufeland, physician to Goethe and founder and editor of the *Journal der practischen Arzneykunde und Wunderarzneykunst*, in which Hahnemann published his first, important article on the Law of Similars in 1796. Hufeland's most significant legacy was his book on macrobiotics entitled *Die Kunst, das menschliche Leben zu verlängern* (*The Art of Prolonging Life*), published in its first version in 1794 and still in print today.[24] In it Hufeland laid down eleven characteristics of *Lebenskraft*, including the body's capacity to react to stimulation from the environment, to fight against destructive influences, such as decay and frostbite, to be strengthened or weakened by external influences, and the presence of *Lebenskraft* within every part of the body. In order to guarantee longevity, *Lebenskraft* needed to be regulated and, in particular, not subjected to excessive physical or mental stimulus, which weakened it and hence curtailed one's life. Thus, in line with eighteenth-century anthropological medicine, for Hufeland regulating *Lebenskraft* involved both moral and physical realms.

Hufeland's (and Hahnemann's) belief in *Lebenskraft* goes hand-in-hand with their empiricism and therapeutic dedication to ameliorating each patient's health. In the context of writing about Barthez's *principe de vie*, Elizabeth Williams points out that

> embedded in the concept of vital force was the idea that spontaneity, variability, and autonomy from the rigid laws of organized matter were the very definition of "life." The capacity of the physician to recognize and gauge variability – in both the patient's fundamental constitution and the myriad external factors that made up the patient's milieu – also lay at the heart of vitalist therapeutics. If treatments were not attuned to the particulars of the patient's experience, they were doomed to failure. (94)

In this respect, Hufeland and Hahnemann follow the Montpellier school in their Hippocratism as well as in adherence to vitalist tenets.

But Hufeland's usage of the term *Lebenskraft* also ran concurrent with the discourse among scientists of his era. For Hufeland too, *Lebenskraft* was an objective reality, although, paradoxically, it was beyond human sight and comprehension: "In this manner has been introduced into physics an infinite variety of powers: the power of gravity, the power of attraction, the electric power, that magnetic power, which, at bottom, signify nothing more than the letter that expresses the unknown quantity in algebra. We must, however, have expression for things whose existence is undeniable though their agency be incomprehensible" (*Art of Prolonging Life* 25). Despite this comparison to other physical forces in nature, Hufeland also writes: "The vital power is the most subtle, the most penetrating, and the most invisible agent of nature, with which we are as yet acquainted. In these respects it exceeds light, electricity, and magnetism, to which, however, it seems to have the closest affinity" (ibid. 26).

Attenuating these earlier claims about the undeniable existence of *Lebenskraft* (despite its invisibility), in 1798 Hufeland published an article, "Mein Begriff von der Lebenskraft" ("My Concept of the Vital Force"), in which he specified that the concept of *Lebenskraft* was merely a cipher or useful philosophical term, in short, a merely heuristic notion that designated the inner, but ultimately unknowable, fount of vitality and regeneration in the body. Hufeland's hesitation seems to be a tacit admission that seeing *Lebenskraft* as a natural phenomenon hides the fact that it is merely a postulate and has actually little to do with "nature." Others were beginning to express their scepticism as well. In his essay on *Lebenskraft* in 1795, the physician Johann Christian Reil opens by saying that such forces have no final, absolute grounding in experience (*Erfahrung*), and hence we cannot ascribe material existence to them ("Von der Lebenskraft" 2). *Lebenskraft* is a subjective concept, the form in which we imagine the connection between cause and effect (ibid. 23). Reil regarded life processes as arising from chemical reactions in the body, not from some indeterminate force. He insisted that, as long as one cannot understand fully the chemical makeup of the body and how heat, electricity, and oxygen work in it, one cannot know how *Lebenskraft* operates (ibid. 15). From a very different arena, too, the Kantian professor of theology and philosophy in Jena, Carl Christian Erhard Schmid (1761–1812), wrote that such a formative power was nothing but an empty fiction outside the realm of experience (1: 129).

Such warnings, however, went largely unheeded for half a century. For instance, Arthur Schopenhauer (1788–1860), in 1836 in *Über den Willen in der Natur* (*On the Will in Nature*), discussed Haller, Treviranus, and writings on mesmerism and plant physiology, searching for where the term *Wille* was deployed because he equated it with a life force: "Das Leben ist Erscheinung des Willens" ("life is the appearance of the will"; 407). Because *Lebenskraft* still had such currency, by the late 1830s the physician and painter Carl Gustav Carus (1789–1869) exasperatedly wrote that physiology was burdened by many abstruse conceptions about life (xi). He deemed the notion of *Kraft* as problematic if regarded as something objective, not merely an operational abstraction (xi). Even the late-nineteenth-century German zoologist Otto Bütschli (1848–1920) used the terms *Lebensenergie* and *vis vitalis*. This long life of the concept of *Lebenskraft* helps to explain why Hahnemann was using it as late as he was, even with increased frequency from the third edition of the *Organon* of 1824 onward.

How, though, did this debate over natural healing powers involve the medical system of the Scotsman John Brown? Hufeland, for one, championed the notion of *Lebenskraft* in reaction to Brunonian theory.

In 1795 he published in the *Allgemeine Litteratur-Zeitung* an essay, "Erste Beurteilung des Brownschen Systems bei seiner Erscheinung in Teutschland" ("The First Assessment of the Brunonian System upon Its Appearance in Germany"), the first of a number of polemics against Brunonianism, which, he charged, lacked respect for the independent self-healing of nature.[25] Although Hufeland stood for a pragmatic and empirically based approach to healing, in opposition to Brown and his followers in Germany, a number of these *Naturphilosophen* also believed in the organic healing powers of nature. For these thinkers, medicine was a speculative science according to which the idea of life was synonymous with the Absolute. Röschlaub, for instance, like Hufeland parting ways with Brown, spoke of a *vis medicatrix naturae*, while Walther wrote on an article, "Über die Heilkraft der Natur"("On the Healing Powers of Nature"), in 1808. For Walther, nature herself was the true artist and physician. Indeed, for several writers of this period, nature was a genius in her capacity to heal; the physician could only aid her in her artistic accomplishments.[26]

Hahnemann joined the voices of Hufeland and others in challenging Brown's disregard of nature. He charged Brown with the "calumniation of nature" ("Three Current Methods of Treatment," *Lesser Writings* 547) because the latter did not trust the powers in nature and believed that they needed to be artificially stimulated or weakened. Hahnemann sarcastically wrote of Brown that "no medical sectarian, *apparently*, knew less about nature than he" (545). Because of the all-pervasive, essential energy that conjoined human and other forms of life, the founder of homeopathy insisted that his science obeyed the code of nature. Already from the opening of his 1796 essay he advocated imitating nature. Then, in the 1810 edition of the *Organon der Heilkunst*, he wrote that nature possesses her own energy ("eigene Energie der Natur"; *Organon-Synopse* 620, §174). Later, in the 1822 edition he refers to nature's healing law ("Natur-Heilgesetz"; "Einleitung," *Organon-Synopse* 178). In "Heilkunde der Erfahrung" he likewise termed homeopathy a "treatment so conformable to nature" ("The Medicine of Experience," *Lesser Writings* 453) as opposed to the "unnecessary artificial disease" arising from medicinal doses that were too strong (455). It was "the highest aim of the reflecting mind," he writes here, to imitate nature, "our great instructress" (469).[27] Thus he wrote: "Nature acts according to eternal laws . . . She loves simplicity, and effects *much* with one remedy, whilst you [physicians] effect *little* with many. Seek to imitate nature!" ("A Preface," *Lesser Writings* 350). Moreover, one could see in the "so remarkably different complaints and sufferings of each single patient . . . this distinct voice of nature" ("Old and New Systems of Medicine," *Lesser Writings* 717). In addition, as with Walther, Hahnemann claimed that nature possessed artistic agency: hers was, in fact, a divine art that allowed the smallest remedy to perform gently and inconspicuously the greatest effects. In brief, nature herself was the best homeopath.

As is to be expected, Hahnemann's use of the term *Lebenskraft* parallels that of other physicians of the era to indicate that power that serves to maintain equilibrium within the body.[28] Like Hufeland, Hahnemann regarded disease as diverging from a "regular state of health" ("Old and New Systems of Medicine," *Lesser Writings* 720). Once homeopathy removes "accidents and symptoms" (ibid.), health is restored. *Lebenskraft* signifies the regenerative, healing force that, in responding to the remedy, overcomes illness. It represents the adaptability of the body and, as such, operates in tandem with nature. Repeatedly in the *Organon*, he refers to the "Gefühl des Lebensprincips," indicating that vitality is something palpably sensed – it is registered as a feeling. Because one's *Lebenskraft* or *Lebensprincip* is a responsive, cohesive entity, once the remedy is administered it will affect the body in whole. At the same time, precisely because it is a unified principle (see *Organon* 116–17, §42) within the entire organism, the concept allows Hahnemann to interpret all symptoms as interconnected. Each and every symptom, however inconspicuous, is a sign of the entire body being affected and in need of rebalance. And because every person's *Lebenskraft* is unique and will react differently, so too are these symptoms diverse (*Die chronischen Krankheiten* 3: vii). That said, although Hahnemann upheld the widespread belief in a *vis medicatrix naturae*, he carried the Hippocratic recommendation that less meant more to its most literal level. Through the infinitesimal dose, Hahnemann developed the belief in the spontaneous, healing powers of nature in a radically new direction.

There are moments in the *Organon* where Hahnemann seems to use the term *Lebenskraft* to contradictory purposes. On the one hand, he refers to it as instinctual, mindless, involuntary, or simply animalistic ("bloß animalisch"; "Einleitung," *Organon-Synopse* 143). Disease is described as the passive suffering of *Lebenskraft*, and it is that which causes the symptoms of illness in the body. Although in itself invisible, *Lebenskraft* manifested its discord ("Verstimmtheit" and "Verstimmung"; *Organon-Synopse* 299, §29) concretely as ailments. In other words, morbid symptoms can be seen as a failed attempt of the organism to overcome what has brought on the illness, in which case Hahnemann calls *Lebenskraft* sick or weak. One's *Lebenskraft* could become even more debilitated because of allopathic treatment. Hahnemann thus tends to use the above adjectives to describe how *Lebenskraft* struggles to operate *without* the assistance of homeopathy. Hence, in the earlier sections of *Organon*, it seems as if Hahnemann is attacking either those physicians, like Hufeland, who in their therapeutic nihilism rely almost exclusively on *Lebenskraft* for natural self-healing, or those physicians, like Brown, who would overprescribe medication.

On the other hand, while the vital life force can be described as being instinctual and mindless, it is also a spiritual power ("geistartig"; *Organon-Synopse* 265, §11).[29] Prior to the third edition of the *Organon*, Hahnemann does not use the term *Lebenskraft* extensively. But, to emphasize nature's independent capacity for agency and self-healing, by 1829 he rephrases *Leben* and *Natur* as *Lebenskraft*, *Lebensprincip*, *Lebens-Energie*, *Lebens-Erhaltungs-Kraft*, and *Autokratie*. The vital force can be retuned ("umstimmen"; *Organon-Synopse* 307, §33) with the assistance of homeopathy.[30] Because *Lebenskraft* is incapable of memory or reason, the homeopathic remedy can trick it into dimming and extinguishing the sensation of *Krankheits-Verstimmung* (*Organon-Synopse* 309, §34). The *Lebenskraft* is merely passive, suffering, and receptive with regards to the primary effect of a drug or disease, yet more energetic in the secondary effect ("Nachwirkung" or "Gegenwirkung"; *Organon-Synopse* 410–11, §63), in other words, in response to the homeopathic, minimal dose. By the fifth and sixth *Organon*, this energetic understanding of *Lebenskraft* led Hahnemann to equate it with *"Dynamis"* (*Organon* 100, §13; 107–8, §29; and 158–9, §117).

In addition, Hahnemann introduces a further variation on the concept of a vital force that sets him apart from Reil and Hufeland. Hahnemann sees it operative not just within the *living* body but within the *inanimate* world as well, ecologically conjoining the two.[31] Hahnemann's understanding of vitalism goes beyond that of the physicians of his time, who restricted their discourses to human health, and approximates the position of the physicists, such as Ritter, who saw an energetic force permeating *all* matter.[32] As well, whereas Hufeland and Reil eventually saw *Lebenskraft* as a regulatory principle, Hahnemann and Ritter saw this energy as an objective reality, that, although beyond human sight, was accessible to the human mind through scientific investigation. Another difference from the physicians of that time was that, whereas the Brunonians claimed that one could heal illness by altering by degree external, *physical forces* of excitement, Hahnemann trusted in the *spirit-like animating force* in the remedy that had the capacity to heal. In sum, in defining this third law of homeopathy, Hahnemann stressed that the botanical substance contained spiritual potency. More than any other physician of the time, Hahnemann thus brought the vitalism of the botanical realm (later in life he was to include the mineral) to bear on the medicinal cure.[33] In his major work, the *Enchiridion medicum* of 1837, Hufeland recognized this aspect of homeopathy when he wrote that "even homeopathy, which considers itself so far above nature, is the best proof of its effective force, for . . ., when it chooses the similarly functioning remedy . . ., the reaction of nature in the same creates that inner natural process of healing to cure the illness" (2–3).

Several instances in Hahnemann's writings give testimony to how he conceived of both the body and the remedy working together dynamically and synergistically. Another way of rephrasing this transmission and mediation would be to call them "processes of feedback," as Robert Mitchell (151) has termed the systems of Hahnemann's contemporaries, the pre-Darwinian biologist Jean-Baptiste Lamarck (1744–1829) and Idealist philosopher of the spirit Hegel. Hahnemann states that, just as external forces affect one's *Lebenskraft* in a dynamic fashion, so too must medicines operate energetically: "*It is only by means of the spiritual influence of a morbific agent, that our spiritual vital power can be diseased; and in like manner, only by the spiritual (dynamic) operation of medicine, that health can be restored*" (*Organon* 101, §15). Again, "the *dynamic* action of medicines, like the vitality itself, by means of which it is reflected upon the organism, is almost *spiritual* in its nature" ("The Medicine of Experience," *Lesser Writings* 466). And later in the second edition of the *Reine Arnzeimittellehre* he contends: "Now because diseases are only dynamic derangements of our health and vital character, they cannot be removed by man otherwise than by means of agents and powers which are also capable of producing dynamical derangements of the human health, that is to say, diseases are cured virtually and dynamically by medicines" ("Spirit of the Homoeopathic Doctrine of Medicine," *Lesser Writings* 620).[34] The invalidated *Lebenskraft* in the individual was to be stimulated via a medicine that concentrated in itself the nonphysical yet natural power. Hereby it is worth noting that the "internal immaterial . . . healing power . . . is so extremely different in every active substance, from that of every other" ("The Sources of the Common Materia Medica," *Lesser Writings* 672). Hahnemann speaks of the "plant or mineral, each differing from the other in its peculiar invisible, internal, essential nature" (676).

Goethe wrote: "Matter never exists without spirit, spirit never without matter" (WA 2.11, 11). Life inheres in matter; and one life force is in all of nature. Translated into homeopathic terms, botanical substances are not inert but essentially alive. The simplest of matter, even on a micro-scale, is active. The homeopathic remedy achieves the materialization of spirit and the spiritualization of matter. The human being, through its responsiveness to the information received from the globuli, can participate in a concatenation of being, an interconnectivity and permutation of things. Matter is one and living. Hahnemann wrote: "Everything in nature lives and is force [*Kraft*]; we must only know how to bring it to life and to develop its power" (*Gesammelte kleine Schriften* 726). His is a dream of universality, one that he shared not only with Goethe but also with the *Naturphilosophen*, such as Lorenz Oken and Schelling, who introduced living spirit into nature. Schelling wrote in 1801: "Nature will no longer be something dead to us . . ., no longer just filling the space, but rather something alive . . . and complete in itself" (*Werke* 2: 735). And Oken enthused: "The spirit is only the impulse of nature, and nature only the animated spirit" (*Lehrbuch der Naturphilosophie* 515). As Dietrich Engelhardt has pointed out, however, the *Naturphilosophen* based proof for the unity of nature on the transference of organic categories onto what cannot be validated ("Fessellos" 39). With Engelhardt's contention in mind, then, how is one to evaluate the scientific, empirical status of such a poetic idea of life? There are two opposing ways of answering this question.

One way is to stress that homeopathy is a movement synchronous with other scientific efforts around 1800 that wanted to provide tangible evidence of the intangible. It can be compared, for instance, to the experiments that made visible the workings of electricity, beginning with Luigi Galvani, who claimed to be the first "to hold in his hands, as it were, this electricity which is concealed in the nerves, and to draw it forth from the nerves and to set it practically before our eyes" (Galvani 79). Galvani thought he had proved the presence of innate animal electricity in the frogs on which he experimented, although, as Volta later challenged, the electricity was actually produced by contact with the metal. Ritter took these trials one step further by conducting the electrical tests on his own body and thus like Hahnemann engaged in self-experimentation. He believed that these currents were a peculiar agent that flowed through everything. For all these scientists, an infinite process undetectable to the naked eye was still able to produce corporeal, physical results, bringing an unknown reality to light. As Iwan Rhys Morus summarized in *When Physics Became King*, in the nineteenth century there was an "increasing array of experiments and instruments designed to make electricity visible" (98).[35]

The Law of Minimum also seems to be indebted to the mathematical laws of calculus and the concept of the infinitesimal. Analogous to a function, $f(x)$, that approaches zero as x approaches infinity (i.e., the distance between $f(x)$ and zero can be made infinitesimally small), according to the function operative in homeopathy's Law of Minimum, the medicinal ingredient could still retain certain properties even though the amount of it was quantitatively negligible. Indeed, Newton referred in his *Principia* to vanishing quantities. Homeopathy, too, offered a theory of how substances could exist even though they were so minuscule that there was no way to see or measure them with the naked eye. That is, homeopathy, like calculus, is indebted to a principle of continuity such that reality, like the line of real numbers, is infinitely divisible (or in Leibniz's words, *natura non facit saltus*). In fact, in the 1833 edition of the *Organon*, Hahnemann himself says: "Mathematicians will inform them, that in whatever number of parts they may divide a substance, each portion still retains a *small share* of the material; that, consequently, the most diminutive part that can be conceived never ceases to be *something*, and can, in no case, be reduced to nothing" (*Organon* 221, §280).

But, the other way of approaching the question of the scientific status of homeopathy is to acknowledge that, for all its indebtedness to Leibniz's mathematics, homeopathy remains an *inversion* of calculus. The strength or energy of a remedy does not diminish, according to Hahnemann, but actually increases exponentially with each dilution. In other words, the result of Hahnemann's setting up rules to govern organic life is ultimately a poeticization of nature. He explains the efficacy of homeopathy by recourse to an infinite spirit not via mathematical measurement. In this respect the laws of homeopathy are not the same as calculable mathematical and physical laws, however much the scaled computation of dilutions lend an air thereof. It thus has not entered the annals of scientific breakthroughs of the era, such as Ritter's discovery of ultraviolet light, that were inspired by the desire to empirically validate vitalistic presuppositions and that Richard Holmes has classically documented in *The Age of Wonder* and Iwan Rhys Morus in *When Physics Became King*.

Still, homeopathy falls fully in line with the Romantic fascination with calculus as investigated recently by the scholars John H. Smith, Howard M. Pollack-Milgate, and David W. Wood. Smith, for instance, scrutinizes how Friedrich Schlegel, in an endeavour to contemplate the infinitely small together with the infinitely large, does so "through the tools of mathematical thinking (not necessarily the rigors of mathematical proof)" ("Friedrich Schlegel's Romantic Calculus" 251). According to Pollack-Milgate, Novalis investigates "the possible presence of the infinite within the finite" (62). And Wood concludes that "there are two kinds of mathematical operations for Novalis, the purely quantitative operations concerned with numerical facts and the measurement of external sensible nature, and a qualitative operation, concerned with the higher self or intelligible inner world of the human being" (270). Similarly, trying to find a "place for the infinite in this most finite of possible worlds that we still inhabit" (Pollack-Milgate 55), Hahnemann imbues incrementally diluted micro-doses (up to the one decillionth, for instance, in the standard C30 potency) with absolute energy and healing powers.

Hegel recognized in *The Science of Logic* (published posthumously in 1832, although the first version of part 1 appeared in 1812) precisely how difficult it was to get this conceptual hold on infinitesimal calculus and to use or apply it: "The *mathematical infinite* is interesting in part because of the expansion of mathematics and the major results which its introduction has produced in it, but also in part because of the oddity that this science has to date still been unable to justify its use conceptually" (5: 279). This hesitation did not stop him, however, from postulating, just as Hahnemann did, the inverse ratio between quantity and quality in *The Science of Logic*. As Hegel put it, each is not only "the negative of the other" (5: 378), but, "to the extent that one increases or decreases, the other likewise increases or decreases and would do so in the same proportion" (5: 450). Hegel, moreover, called this "*double transition*" (5: 384) the "ratio of powers [*Potenzenverhältnis*]." Be it in Novalis, Schlegel, Hegel, or Hahnemann, then, the new paradigm in mathematics opened onto excitingly new figurative applications.

We are now in a position to situate the *Denkstruktur* with which homeopathy operates within the broader development of philosophic and poetic thought around 1800, starting with Kant. Hahnemann considered himself as Kantian. In a letter to Kant's French translator Charles Viller in 1811, he wrote: "I respect Kant very much, especially because he drew the limits of where philosophy and all human knowledge end in experience" (qtd. in Tischner, *Geschichte der Homöopathie* 229).[36] The philosopher, though, "drew the limits" quite differently with regards to the vital force in the organism. In the *Third Critique*, where he discusses the teleological or purposive organization of natural beings, Kant acknowledges the necessity of positing an "self-formative force [*in sich bildende Kraft*] . . . which cannot be explained solely by the capacity for movement (mechanism)" (5: 486). But he also calls this force that synthesizes the parts of an organism an "inscrutable property" (5: 486); it serves merely as analogous to life (5: 487). It must remain a regulatory concept, not constitutive of being (5: 487). As Joan Steigerwald has explained: "Kant's critical examination of our teleological judgments of organisms thus introduced a distinct mode of judgment, . . . a mode of judgment with only subjective, not objective validity" ("Rethinking Organic Vitality" 57).[37] In other words, we can imagine the internal purpose of such a driving force but, because of the limitations of our faculties, we cannot say more about its cause. Kant posits an "intuitive understanding" (5: 524) that allows us to imagine the connectivity of the parts of nature. An intuitive understanding, however, does not allow us to prove this *Kraft*, that is, to find the ground of it. To be sure, Kant acknowledges that an organized being is not merely a machine: but we do have to rely as much as possible on mechanistic explanations of the bodies working, otherwise we engage in lazy reasoning.

Hahnemann, too, says there is more to life than can be explained by iatromechanical views of the body. But, unlike Kant, he refuses to acknowledge when anatomical and physiological explanations of natural processes are appropriate. Instead, in order to explain the entire body's functioning, he has recourse to the existence of *Lebenskraft*, and not merely as a heuristically operative notion, as Hufeland cautioned in 1798. That is to say, in order for the homeopathic remedy to be effective, it must trigger the responsiveness of some entity, which was *Lebenskraft*. Moreover, the remedy itself was not inert matter but governed by dynamic forces. In this respect, Hahnemann goes far beyond Kant: these forces are no longer merely regulatory, he claimed, but demonstrable. Indeed, they had to be evincible in order to prove that homeopathy healed. Kant, by contrast, cautioned that even someone with Newton's status could never explain how something as simple as a blade of grass could, according to natural laws, come into being (5: 516).

Hahnemann's leap into the spiritualization of the infinitesimal dose is a counterintuitive, utopic reversal on a par with Novalis speculating at the time that the Golden Age was imminent and that poetry could lead us to it.[38] Many other thinkers of the period were eager to overstep the divisions that Kant had erected. In fact, precisely because they were acutely aware of the inability to know the inner workings of nature beyond appearances, they aspired to formulate a meta-critical position that would unify mind and nature. For instance, developing the Königsberg philosopher's concept of intuitive understanding in a direction he would not have approved, Goethe relied on intuitive capacity to link similarities he sought in living organisms and therefore to envisage a developmental unity to them.[39] Schelling and Oken, too, wanted to transcend Kant's hesitation, even the position of earlier vitalists, such as the naturalists Treviranus and Blumenbach or the physicians Hufeland and Reil, who restricted their investigations to living forms. As Sibille Mischer has pointed out, unlike these earlier theoreticians, Schelling and his followers declined to impute forces solely to the organic realm (7–8). This is not to say that the Romantics were not completely open to the cardinal methods of observation and experimentation; they were. But they also associated with the terms *Kraft* and *Materie* (matter) something very different, namely, a thoroughly developed metaphysical concept (Mischer 34). Theresa Kelley has expressed this progression away from Kant as follows:

> *Naturphilosophie* writers transformed Kant's regulatory principle that we ought to proceed as if organisms have inner purposiveness into a constitutive claim about what nature is. This swerve from Kant decisively set *Naturphilosophie* on its own philosophical course. Not content to suppose that some organic principle is at work alongside mechanical processes of the kind being discovered in chemistry, physics, and biological life, Friedrich Schelling and then others insisted further that mechanical processes are themselves directed and supervised by an organic spirit and, still further, that the difference between spirit or mind and nature or matter is one of degree not essential kind. (220)

Despite his own professed allegiance to Kant's empiricism, Hahnemann must be included among these later thinkers who desired to surpass the limits of cognition set up by Kant by stepping from the material into the immaterial world.[40]

Homeopathy and Mesmerism

With an understanding of how central a concept *Lebenskraft* is to Hahnemann, it is now possible to compare homeopathy to two other medical practices that arose at approximately the same time and that share with homeopathy an afterlife until today, although they too were hotly contested for several decades around 1800. One of these, mesmerism, originated with Franz Anton Mesmer's (1734–1815) theory of "animal magnetism."[41] At first, it would seem to have little in common with homeopathy because it is based not on the dispensing of remedies but on influencing the flows of magnetic forces within the body through hypnotism, the use of magnetic rods, and the gliding of hands over a patient. The other modality, vaccination with the cowpox virus, by contrast, shares basic concepts with homeopathy – the Law of Similars and the Law of Minimum. In vaccination, a minimal amount of a similar illness has, though not curative, preventative effects. But whereas mesmerism proves to resemble homeopathy in many ways, vaccination turns out to be essentially different, as can be illustrated with reference to how *Lebenskraft* operates in each.

Hahnemann certainly did not follow Kant with respect to mesmerism, however much he otherwise claimed (falsely, as we have just seen) to follow the epistemological limits Kant set down for reason. The Königsberg philosopher derided mesmerism and put it on a par with alchemy and ventriloquism (6: 441).[42] Hahnemann, on the contrary, was a great proponent. His protocol book (G2) indicates that he tested magnets on himself. It also includes excerpts from Ritter and Eberhard Gmelin (1751–2809) on galvanism and mesmerism. The first homeopath often referred patients to a mesmerist, as his case journal of 1830 illustrates (120–1). From the third (1824) through the sixth (1842) version of the *Organon* he includes an addendum on this controversial healing modality. Mesmerism, as a form of energy work, underscores and illuminates certain salient features of homeopathy. Like the homeopath, the practitioner of animal magnetism manipulates *Lebenskraft*, that is to say, rights the dynamic flows in the body. Following his disciple Ernst Stapf, who in 1823 published an essay entitled "Zoo-Magnetic Fragments," Hahnemann highlights two ways in which mesmerism works in tandem with homeopathy. First, he claims that mesmerism "acts [like homeopathy] by imparting a uniform degree of vital power [*Lebenskraft*] to the organism when there is an excess of it at one point and a deficiency at another" (*Organon* 227, §293). He continues by saying that mesmerism "acts by immediately communicating a degree of vital power to a weak part or to the entire organism . . . without interfering with the other medical treatment" (*Organon* 228, §293). Stapf called this equalizing *Lebenskraft* the dietetic application of mesmerism. It was a kind of "supplement for the nerves" (5) that the powerful magnetizer could impart to the weaker, passive individual.[43] It could be used to counteract the loss of appetite, persistent constipation, sleeplessness, paleness, depression, or nervous exhaustion caused by excessive intellectual activity (8). Stapf recommended that a patient seek out a mesmerist first in order to balance the system enough that the homeopathic remedy could work effectively (10).

As early as 1805 Hahnemann notes how "the heroic power of *animalism* (animal magnetism) . . . displays such an energetic action on very sensitive, delicately formed persons of both sexes, who are disposed either to violent mental emotions or to great irritability of the muscular fibers." He therefore cautioned that it could act "with more than excessive violence in those states of morbid sensibility and irritability" ("The Medicine of Experience," *Lesser Writings* 464). Stapf similarly argued that, when the energy imparted to a sensitive person was too powerful, mesmerism would induce a crisis-like effect. Consequently, he called for its measured, that is, homeopathic dose. But how much was appropriate? The powers of the mesmerist had to be penetrating enough to gauge the quantity of the effect so as to avoid overpowering the patient. If necessary, the treatment needed to be repeated. The same ineluctable problem of minimalizing and gauging the frequency of dosage preoccupied Hahnemann throughout his career, as we saw earlier in this chapter.

The second way in which mesmerism could work as a supplemental therapy to be integrated into the homeopath's arsenal derived from the Law of Minimum. Hahnemann's and Stapf's argument was that, just as a medicine in too strong doses can be deleterious and the homeopathic dose remedial, so too the mesmerist must avoid crushing a weak individual (the examples of which are numerous in the literature on mesmerism) and only gently produce symptoms of excitation, similar to the disease, yet calculated enough to cure it. Hahnemann writes: "The powerful will of a well-intentioned individual influences the body of the patient by the touch, [and] acts homeopathically by exciting symptoms analogous to those of the malady – and this object is attained by a single transit, . . . gliding the hand slowly over the body from the crown of the head to the soles of the feet" (*Organon* 227, §293). In mesmerism, too, the Law of Minimum works in tandem with the Law of Similars.

Both mesmerism and homeopathy thus involve sympathy, flows, and exchange between and across bodies and matter. But there are other resemblances. In addition to restoring health by revitalizing *Lebenskraft*, both mesmerism and homeopathy similarly demand a receptive, open patient and respond to the capacity in the individual to be stimulated. Both call in various ways upon the invisibility topos: like the dynamic force in the homeopathic remedy, in mesmerism what heals cannot be seen yet acts upon the body. The movements of the magnet, a pendulum, or the mesmerizer's hands are governed by natural phenomena beyond conscious will that make manifest an eternal but invisible vital principle. As one thinker of the period observed in 1817: "Animal magnetism offers the soul a view into its secret laboratory, and we happen on discoveries that we are denied in the ordinary course of things. Animal magnetism is an experiment that we conduct with the organ of the soul to study its inward nature, much in the same way the natural scientist conducts experiments to investigate the inner properties of the physical matter" (Eschenmayer, 33). Like homeopathy, then, animal magnetism makes manifest forces otherwise hidden from human eyes. Indeed, James Frazer in *The Golden Bough* classifies "Homeopathic or Imitative Magic" under the category of "Sympathetic Magic," according to which "things act on each other at a distance through a secret sympathy, the impulse being transmitted from one to the other by means of what we may conceive as a kind of invisible ether" (14).

Gotthilf Heinrich Schubert (1780–1860) was one of the most important Romanticists to write on animal magnetism in both *Die Symbolik des Traumes* (*Symbolism of Dreams*) and *Ansichten von der Nachtseite der Naturwissenschaft* (*Perspectives on the Dark Side of the Natural Sciences*). Although he does not refer to homeopathy, there are moments in his work that evoke this other unconventional medical practice. He points out how pure gold, similar to the touch of the mesmerizer, can give a pleasant feeling to the person in a magnetic slumber (*Ansichten von der Nachtseite* 336). In 1819, Hahnemann recommended "the smallest dose of pulverized gold attenuated to the billioneth degree" ("On the Uncharitableness towards Suicides," *Lesser Writings* 695) to remove suicidal thoughts. Schubert notes the properties of magnetized water (336), analogous to how, in homeopathic dilution, the menstrum carries and transmits energetic properties. Overall, the reason that mesmerism culminates Schubert's discussion of natural science is that it gives the perfect example of "the eternally harmonic working together of the World-All in every part" (372). The goal of his study is to point to "the vital soul, which descending from above, suffuses all of nature even into the most external and smallest elements" (372). Or, as Stapf phrased it in conclusion to his essay, what drives the "thousand-fold metamorphoses in life" is the "magic wand" of the "dynamic, the virtual" (27–8).

In this respect, the rhetoric Stapf adopts is telling for its ambiguity. He speaks of life processes as obeying a "magic wand." The effect of animal magnetism could be either "material" or "purely spiritual" (3). Its mystery is revealed to both the "deeply perceiving eye of the serious investigator, as well as the daring wings of fantasy" (3). It follows "the eternal laws of nature" despite the "hieroglyphs" and "fragments of the great, mysterious Temple of Isis" (3–4). In other words, Stapf considered mesmerism to be a phenomenon that hovered between a natural and the supernatural explanation. It saw forces at work that could not be materially explained. Its power, if you will, was its ambiguity. In this it fully resembled homeopathy.

Hahnemann, too, insisted that the phenomenon of homeopathic healing could not be denied, although its process escaped comprehension. It played with the borders between psyche and physis. Like mesmerism, homeopathy was surrounded by an air of the strange and mysterious, yet both claimed to be fully in tune with nature. The air of empirical scientific observation and investigation lent proof to the spiritual. Hahnemann's writings extensively narrated and documented bodily sensations; he was immersed in the sensuousness of life processes. Yet at the same time he toyed with the inexplicable. The magical, fairy-tale-like effect of the globuli laid claim to being grounded in reality. In writing about the tales of the Romantic fantastic, as in the stories of E.T.A. Hoffmann (1776–1822), such as "Der goldne Topf" ("The Golden Pot"), French literary critic Tzvetan Todorov (1939–) observed that it oscillated between such diametrically opposed possibilities. In the genre of the fantastic, the reader could not tell whether the events narrated belonged purely to the realm of the supernatural or whether they could be given a plausible, realistic explanation. In the latter case, the bizarre occurrences could be attributed to the perspective, psychological imaginations, and convictions of the main protagonists. This suspension between two interpretations was never resolved in the course of the fantastic narrative. An explanation of the workings of homeopathy, too, hovers between the spiritual, on the one hand, and the material, on the other.

Homeopathy versus Vaccination

Where homeopathy and mesmerism differed, of course, was that homeopathy never attracted the same degree of legendary, sensationalist attention. Mesmerism was a melodramatic practice with charismatic male practitioners and swooning female patients, as depicted, for instance, in E.T.A. Hoffmann's "Der Magnetiseur" ("The Mesmerist"), another example of the fantastic tale where the uncanny, supernatural occurrences strongly invite psychological interpretation. A more serious, new medical practice, however, was gaining ground at roughly the same time. It too divided the medical community and caused no small degree of consternation in the lay population. This revolutionary practice was that of vaccination. In 1796, Edward Jenner (1749–1823), the "father of immunology," introduced vaccination with the cowpox virus,[44] a nonlethal, safe alternative to inoculation (variolation) with the smallpox virus. By 1800, Jenner's work had been published in Germany, and vaccination was soon recommended on a widespread basis, despite pervasive fears of contamination by a nonhuman substance.[45]

As Cornelia Zumbusch reminds us, although the discourse existed around 1800 of being inoculated with hardship, fate, or a virus, there was no medical concept of immunity as we know it today. One could be free *from* contagion, but not immune *against* or resistant to something (10). How, then, was vaccination seen to work? Richard Squirrell writes in 1805 in "Observations Addressed to the Public in General on the Cow-Pox": "Owing to the very slender affinity that the blood has to the cowpox virus, the animal economy, after being compelled to receive it, endeavors by its own laws to evacuate it out of the habit, and which, sooner or later, according to the power of irritability, or the preserving principle, it will, no doubt, accomplish" (19). Here, as in homeopathy, a power or "preserving principle" erases or "evacuates" the original disease or virus. But how the homeopath conceives the actions of such a vital principle differs significantly, as can be demonstrated by reference to the main proponent at the time on how to inoculate oneself via exposure to hardship, namely, Hufeland.

There are moments in Hufeland's *Die Kunst, das menschliche Leben zu verlängern* that seem to resemble in its call for moderation Hahnemann's notion that a toxin in small doses serves to reinvigorate *Lebenskraft*. For instance, Hufeland believes that exposure of the body to moderate degrees of cold can strengthen or toughen it: "By such daily enjoyment of air, acquainted and familiar with the free atmosphere . . . people are thus secured against one of the greatest evils that usually afflict mankind, I mean *too much sensibility in regard to all impressions and variations of the weather*. This is one of the most abundant sources of disease; and there is no other method of counteracting it, but to harden oneself by daily exposure to the open air" (*Art of Prolonging Life* 259). Towards the end of this essay Hufeland speaks of how to acclimatize oneself to grave dangers to the body: he recommends "the utmost possible agility and readiness in all bodily exercises" (315), and that one should "endeavor to render the mind intrepid; to give it strength and philosophical equanimity; and accustom it to sudden and unexpected events . . . One will thereby guard against the physical injury of sudden and alarming impressions" (315–16.) To get a little less sleep than normal, to drink a glass more of wine, or to do gymnastic exercises, keeps the body from residing in too passive a state and hence serve as preventative measures against disease (303). In other words, they offer examples of "like *preventing* like." Although Hufeland himself was later to recommend against the smallpox vaccination because it represented to him an invasion of the body's integrity, his notion of steeling *Lebenskraft* parallels the notion of guarding and protecting the self that inoculation and/or vaccination represent: a small dose of a poison will boost one in the face of a dangerous exposure. It is a poison with a curative, freeing role – a pharmakon. In sum, for Hufeland, *Lebenskraft* is best sustained and guaranteed equilibrium by the moderation of stimuli. As a final example, for the elderly, who see a decrease in the vital life force, he recommends a moderate degree of wine for stimulation. Yet in a paradoxical example of how a diminished *Lebenskraft* actually sustains and prolongs life, he also says that, because the intensity of the elderly's responsiveness is reduced, they are also less susceptible to infectious diseases than the young (ibid. 322).

Although Hahnemann also prescribes moderation and explicitly states that vaccination operates according to the principle of like curing like ("Cure and Prevention of Scarlet-fever," *Lesser Writings* 370; "Necessity of a Regeneration of Medicine," *Lesser Writings* 520; *Organon* 118–20, §46),[46] his notion of gentle recalibration actually veers from Hufeland's defensive hardening of *Lebenskraft*. But first, let me explain how Hahnemann does resemble Hufeland in his early writings on the topic of "Making the Body Hardy," as he entitled a segment in his 1792 book *Freund der Gesundheit* ("The Friend of Health, Part 1," *Lesser Writings* 191). Here he recommends taking precautionary or prophylactic measures that resemble inoculation in order to prevent susceptibility to infectious disease. For example, clergymen and physicians who start to visit a sickbed should "see their patients more frequently, but each time stay beside them as short a time as possible" (ibid. 169). "As in the case in accustoming ourselves to everything, *the advance from one extreme to the other must be made with the utmost caution, and by very small degrees*" (168–9). Exposure to wind and extremes of temperature must be made "always only gradually, interruptedly and by progressive advances" (197). Moreover, "the hardening of the human creature in respect to heat and cold no doubt is commenced with greater safety in childhood . . ., but we require to exercise the greatest caution at first with these tender creatures" (194). In short, the trope found throughout Hahnemann's writings of gentle, minimal alteration finds its origins here in 1792.

Of course, the main distinction between homeopathy and vaccination is that the former works curatively, the latter prophylactically. But in 1801, where he already refers to the "dynamic" action of medicine ("On the Power of Small Doses of Medicine," *Lesser Writings* 387) and to infinitesimal dilutions ("1/432,000th part of a grain of the extract"; "Cure and Prevention of Scarlet-fever," *Lesser Writings* 379), Hahnemann does indicate that homeopathic doses of *belladonna* can be effective in preventing the onset of scarlet fever in addition to curing it. The treatment is both *"capable of maintaining the healthy uninfectable"* and *"given at the period when the symptoms indicative of the invasion of the disease occurs, stifles the fever in its very birth"* (ibid. 377). "Moreover, [it] is more efficacious than other known medicaments in removing the greater part of the *after-sufferings* following scarletina that has run its natural course, which are often worse than the disease itself" (ibid.). To the same effect of linking prophylaxis and cure, Hahnemann writes in 1803 that "there cannot be any *prophylactic* of hydrophobia [rabies], that does not prove itself to be at the same time a really efficacious *remedy for the fully developed hydrophobia*" ("On a Proposed Remedy for Hydrophobia," *Lesser Writings* 390).

But more important than this conceptual linking of inoculative prophylaxis and homeopathic cure is actually the difference between them.[47] The complex dichotomy works as follows: moderation, although also a goal in itself for Hahnemann, is a basic prerequisite to homeopathy's proper functioning. In other words, the body will only be sensitive to the homeopathic remedy if it is not exposed to other stimuli such as wine or coffee. Even early in his career, Hahnemann cautioned against "coarse stimulants and aphrodisiacal arts" that would blunt the senses ("The Friend of Health, Part II," *Lesser Writings* 230). What preserving sensitivity entails is that, rather than toughening one's *Lebenskraft*, homeopathy operates on the basis of increased receptivity to the gentle stimulus of the micro-dose. To be sure, homeopathy is anxious about hygiene and overdose and thereby seeks the least minimal contact with a tangible world. But this desire for purity aims to accelerate invisible flows of energetic forces. This energy was so delicate that by 1833 Hahnemann cautioned that homeopathic preparations were disturbed even by travel. To state the difference concisely, inoculation means inuring, hardening, or habituating oneself precisely in order to avoid the susceptibility that is the precondition for homeopathy's effectiveness.[48] Hahnemann, for instance, claimed that the sick patient is far more impressionable and susceptible to the action of a medicine than is a healthy person ("The Medicine of Experience," *Lesser Writings* 464). The reason is that "in disease the preservative power [*Erhaltungstrieb*] . . . is much more excitable than in health" ("On the Power of Small Doses of Medicine," *Lesser Writings* 387–8). "In a word, all the powers [*Kräfte*] . . . are infinitely more excited in disease" (388). In health, the body stands "in no need of such anxious guardians" (ibid.). Because of this excitability, a patient who is sick with an acute fever doesn't even need to ingest a drop of medicine, for the mere smell of it is enough to produce results (ibid.).

All told, if homeopathy presumes and enhances the integrity of the body, at the same time this body is a permeable membrane. Homeopathy avoids any poisoning that allopathic medicine might generate; it also steers clear of the spectre of contamination by a nonhuman substance (actually, from the instruments used) attributable to vaccination. In terms of comparison to other major physicians of the day, Hahnemann does overlap with Brown and Hufeland to the extent that they all offer variations on the same general eighteenth-century framework of *Reiz*, and they all believe health resides in maintaining equilibrium. But Hahnemann leaves the discourse of irritability, as Haller and Brown use it, with its mechanistic implications, to conceptualize a spiritual and dynamic responsiveness of the body. In addition, Brown, who never relied on the natural powers of healing (*Naturheilkraft*), would not have seen the body as self-regulating to the extent that Hahnemann and Hufeland did. Brown would have sought either to depress the sthenic state or boost an asthenic condition, as he called them, by using powerful stimulants. By contrast, should *Lebenskraft* be in a state of emergency, according to Hahnemann, it demanded gentle, calibrated treatment that would prompt the body to self-attunement. Homeopathy, in short, restores balance micro-instrumentally. That Hahnemann frequently changed remedies can, in fact, be explained by his desire to ever adjust equilibrium.

In "Über die ästhetische Erziehung des Menschen" ("On the Aesthetic Education of Man") Schiller indicates how theatre rallies the "strong side" of man through "inoculation with unavoidable fate" (21: 51). In the scholarly work on the cultural implications of vaccination around 1800, it has frequently been pointed out, the model of hardening is reflected in a discourse – prevalent in the Classical literature of Schiller and Goethe – of moderation, stoicism, impassiveness, endurance, and restraint of affect.[49] These studies, narrowing their focus on the medical history of contamination, infectious illness, and their treatment, do not mention homeopathy, and understandably so, for the latter offers a different model of human communication and understanding of individuality, one in tune more with Romanticism than with Classicism. For one, Hahnemann's attention to the minute variations in *Gemütsstimmungen* in the anamnesis (as discussed in the previous chapter) meant that he encouraged the verbalization of emotion rather than its restraint. For another, homeopathy parallels the Romantic discourse of *Empfänglichkeit* (receptivity, susceptibility, impressionability) and universal sympathy. Novalis prescribes the ideal poetic nature as capable of a many-sided impressionability ("eine vielseitge Empfänglichkeit"; 1: 385). He depicts his hero Heinrich von Ofterdingen as a passive figure, yet openly receptive to the world.[50] Similarly, in his celebrated public lecture of 1794, "Ueber die Bestimmung des Gelehrten" ("On the Vocation of the Scholar"), Johann Gottlieb Fichte (1762–1814) prioritized the concept of *Empfänglichkeit*, listing it as one of the main characteristics of the educated individual (6: 330). And Friedrich Hölderlin (1770–1843) wrote of "die Empfänglichkeit in uns, / Die uns vereinigte mit andern Geistern" ("the receptivity in us / Which unites us with other spirits"; 3: 191).

Finally, situating Hahnemann within the prevalent cultural discourses of his time calls for a contrast between the sublime and the calm – between immunopathic defensiveness and homeopathic pliability. Homeopathy proposes the force field, if you will, of the infinitely small, in contrast to the magnitude of the sublime that Kant discussed in his Third Critique, although, to be sure, both Hahnemann and Kant locate the infinite in nature. In a defensive move – Cornelia Zumbusch calls it an *Abwehrbewegung* (84) – Kant claims that reason reasserts its command over nature when nature in its grandeur overwhelms the senses. Reason redirects the awe back to itself. The mild actions of homeopathy are less in accord with this late-eighteenth-century Kantian sublime than with what in 1852 the Austrian Biedermeier poet Adalbert Stifter (1805–68) famously called the gentle law (*das sanfte Gesetz*) that should guide human nature. Indeed, the poet Bettina von Arnim wrote in 1838 that homeopathy healed "by gently embracing nature" (Schultz 47).

In sum, Hahnemann sees the human body as an open system in a continuous change of energies. The body responds to the vitality of the natural remedy rather than build, as one might explain today, antibodies against an entity. Whereas immunological identity, as inherited from the late eighteenth century, sets life against life, Hahnemann engages life on behalf of life. In this he resembles newer concepts of immunity that speak of auto-reactivity in terms of interaction and of homeostasis as constant regulation of the body's balance. The biologist and founder of general systems theory, Ludwig von Bertalanffy, wrote in 1963: "The organism is not a static system closed to the outside and always containing the identical components; it is an open system in a quasi-steady state, maintained constant in its mass relations in a continuous change of component material and energies, in which material continually enters from, and leaves into, the outside environment" (8). More recently, the French philosopher of science Thomas Pradeu has postulated that "grafts, fetomaternal tolerance, and the tolerance of microorganisms and macroorganisms call into question the . . . claim of the self-nonself theory, according to which the immune system triggers the rejection response against all 'nonself' . . . It is a flawed vision of the organism as 'pure' – perfectly homogenous and endogenously constructed – that has led to the idea that an organism has to reject any 'foreign' entity" (127). It may be helpful to conceptualize the biological individuality that classical homeopathy presupposes along similar lines. The body is capable of adapting to the minuscule stimulus offered by the remedy and thereby auto-restores equilibrium. Or, to put it differently, the idea of the organism that current immunological theory posits can conceivably look back to German Romanticism for a conceptual predecessor.

A Final Note on *Potenzierung*

In his book on Johann Wilhelm Ritter, Walter D. Wetzels writes of the risky self-experimentation that Ritter conducted on himself (101–2). Ritter reports how he would look into the sun for 15 to 20 minutes, gaze at a blue-coloured paper, and discover that it suddenly appeared red. The inversion into the opposite colour, attained, in Ritter's words, through the *Extremisierung* of experience, is at the same time a *Potenzierung* or potentization, for Ritter describes the achieved colour as more intense. Wetzels compares this reversal of normal reality into its heightened counterpart to Novalis's famous definition of Romanticism as sheer qualitative potentization ("Romantisiren ist nichts, als eine qualit[ative] Potenzirung"; 2: 334, #105). He also cites Friedrich Schlegel's dictum that "each development, every growth and advancement has one aim, an ultimate goal, that either necessitates returning to the beginning or leaping into its antithesis" (*Kritische Ausgabe* 12: 434). John Neubauer called *Potenzierung* as conceived by the early German Romantics "development from one level to another by leaps and bounds [sprunghaften Entwicklung]" ("Zwischen Natur und mathematischer Abstraktion" 178).[51]

Today one designates homeopathic preparations according to their "potencies," for instance, as 6X or 30C. But Hahnemann only settled on the term *Potenzierung* rather late in his career. Up until 1821, that is to say, volume 6 of the first edition of the *Reine Arnzeimittellehre*, Hahnemann referred to his preparations as a dilution (*Verdünnung*) or a mixture (*Mischung*). Only thereafter does he deploy the notion of *Potenzierung* and start referring to dynamization. In fact, in volume 6 of the second edition of the *Reine Arnzeimittellehre*, he speculated that potentization can occur infinitely.[52] In §§269 and 270 of the final edition of the *Organon* of 1842, where Hahnemann delineates how to prepare the homeopathic remedies, he stresses that *Potenzierung* is not the same as dilution: "Homeopathic medicinal potencies are still regarded as dilutions, although they are actually the opposite. They unlock the natural substances, they disclose and reveal the specific medicinal forces that are concealed in their inner essence" (*Organon der Heilkunst* §269, 192, note 4).[53] He sets up a tidy equation: less of the original matter (*das Materielle*) relates precisely to increased powers (*Kräftigkeit*, §270, 195, note 7). Conversely, *Potenzierung* implies that matter, in and of itself, has correspondingly less spirit: "In its raw state it can only be considered an unrefined, spirit-like being" (§270, 196, note 7). Matter, however, is capable of this magical, spiritual transformation. It becomes "purified and transformed entirely into a spirit-like medicinal force through steadily increasing dynamizations" (§270, 193). "Matter will thus become spiritual, if one may put it this way" (§169, 191).

Wetzels's analysis of Ritter helps clarify why this final choice of *Potenzierung* is an important one for Hahnemann. The term, borrowed from infinitesimal calculus, captures well the unexpected, striking outcome that Hahnemann sought to achieve: by thinning a remedy to the extreme, instead of weakening its effect, its power is increased limitlessly.

This process of transmutation and systematic reversal of matter into spirit occurs with computational precision. But there is more: Novalis and Friedrich Schlegel associated *Potenzierung* with a striving towards the infinite. They deployed the term in some of their most famous characterizations of the Romantic project. Schlegel defined transcendental poetry as potentized into infinity ("ins *Unendliche* . . . potenziert"; *Literary Notebooks* 82, #698). He prescribed in his renowned Athenäum fragment #116 that Romantic progressive *Universalpoesie* should "hover on the wings of poetic reflection, always elevating this reflection [immer wieder potenzieren] and multiplying it, as in an endless succession of mirrors" (2: 182–3). *Potenzierung* designated the flight of the imagination, the journey into the realm of the spirit, as well as limitless progression and growth. It endowed the ordinary, in Novalis's words, with "elevated meaning" (2: 334, #105). The poet of the blue flower trenchantly wrote: "The spirit is the potentiating principle" (2: 516, #243). Schlegel declared that it was the tendency of the human spirit "to aim for higher and even higher potencies" (*Kritische Ausgabe* 8: 382).

In his book *Novalis and Mathematics*, Martin Dyck has noted that the Romantics projected through the use of *Potenzierung* "mathematical thought patterns onto literary criticism" (85). Novalis believed mathematics, capable of infinite perfection, was the main proof of the sympathy and identity of *Natur* and *Gemüth* (2: 838, #422). With this cultural background in mind – where poetics and the sciences were not yet split as they are today – it is understandable why Hahnemann had recourse to the same vocabulary of *Potenzierung* to designate his systematic dilutions. The ascending, ever more spiritual powers of consecutive dilutions retuned the *Gemüth*, now in harmony, sympathy, and oneness with *Natur*. This belief, one can conclude, is equally a "projection of mathematical thought patterns." In fact, Hahnemann may be said to be more Romantic than the Romantics. With its creed of the computable effect of *Potenzierung*, homeopathy trusts that it realizes what Novalis merely envisioned, that is to say, in the poet's words, the "higher art" of *Infinitesimalmedicin* (2: 550, #399).

CONCLUSION

Novalis, the prime spokesperson for the Romantic longing for harmony with nature, mused:[1] "Does not all nature, including the face and the gestures, the pulse and the colours, express the condition of . . . humans? Does not the rock become curiously familiar when I address it? How am I different from the river when I gaze with sad longing into its waves, and my thoughts are lost in its flow?" (1: 224). Novalis is not alone in desiring to see human life and the life of the mind as one with nature. Other voices belong to Samuel Taylor Coleridge (1772–1834), Goethe, Ritter, Oken, and Schelling. Following in the footsteps of Herder's championing of Spinoza, they wove Spinoza-inflected concepts of *hen kai pan*, *natura naturans*, and a *scientia intuitiva* into the fabric of their thought. These concepts will be rehearsed here in conclusion. But whereas one can say that Romantic thought – be it in literature, philosophy, or science – was past its zenith by 1815, homeopathy was only then beginning to gain ground. Given that this medical therapy is alive and well today, it is not an exaggeration to say that Hahnemann was the one thinker among this group to truly bequeath to the twenty-first century the Romantic yearning for a spiritualized nature. Homeopathy maintains its attraction more than ever for people in search of a holistic mode of living within an environmentally troubled world.

In response to the eighteenth-century rational classification of species, the Romantics envisaged a dynamic harmony and web of connections in nature that could be intuited. In terms of the broader history of the connection between science and art, Pierre Hadot, in *The Veil of Isis: An Essay on the History of the Idea of Nature*, has argued that this "Orphic tradition" has been pushed aside in the ascendency of a technologizing, instrumentalizing view of nature. The Romantic "Orphic tradition" offers an alternative interaction, one in which the musical or poetic metaphor of harmony characterizes the relationship to the natural world. Odo Marquard has similarly argued that Schelling's dynamic, organic *Naturphilosophie* was directed against hard sciences based on quantifiable measurement, and that for him aesthetics in its holism offered an answer to scientific compartmentalization. As the nineteenth century progressed, however, the precariousness of an aesthetic response led to its waning, a development hastened by the ascendant view of nature as motivated by drives and the competition between species, a view that sanctioned the human ambition to conquer nature. Consequently, within this wide sweep of Western intellectual history, traces of the "Orphic tradition" have become largely invisible in twenty-first-century culture – except perhaps in the Romantic legacy of homeopathy and its non-technologized approach to wellness. Hahnemann saw nature as giving man through homeopathy the means to health and wholeness. It was a "gentle, safe healing art so consonant with nature" ("Allopathy: A Word of Warning to All Sick Persons," *Lesser Writings* 752).

To call homeopathy "natural" is not merely to state that it is a non-invasive therapeutic intervention, which today is the sense in which it is used and integrated into various naturopathic healing modalities. It is not merely that homeopathy is plant-based rather than pharmaceutically developed in the laboratory. Nor that it engages the body's own curative forces. More importantly, homeopathy is "natural" because it is grounded in a belief in the dynamic activation of a force within nature herself and transmitted by the plant, although enhanced by sequential dilutions, trituration, and shaking. This dynamism, this natural force, is one whose laws Hahnemann claims to have harnessed. He wrote: "Beneficent nature shews us, in the homeopathic method of treatment . . . the most unfailing cure" ("Old and New Systems of Medicine," *Lesser Writings* 723). Bettina von Arnim praised the "excellent and simple effects" of homeopathy, for they sacredly respected nature (Schultz 56). Bönninghausen called homeopathy a child of nature (*Die Homöopathie* 13). Hahnemann himself considered his *materia medica* a codex of nature (*Organon* 169, §143) and that homeopathy followed the healing law of nature ("Naturheilgesetz" [*Organon* 107, §28]). To be sure, the roots of Hahnemann's belief in a beneficent nature go back to eighteenth-century notions of perfectibility and natural order. But these ideas feed, above all via Herder, into the notion of a spirit-like *Kraft* in the natural world, whose laws of operation the Romantics sought to discover. Whether Novalis, Ritter, Oken, Alexander von Humboldt, Goethe, or Hahnemann, they all aspired to unlock signification in nature, above all via the keys of analogical reasoning. A belief in vibrant powers subtending both animate and inert creation, man and nature, pervades their intellectual endeavours around 1800. As the Novalis scholar Gabriele Rommel summarizes, the Romantics link "man's reintegration as a 'natural being' with the idea of an infinitely positive force" ("Romanticism and Natural Science" 216).

Take, for example, the beginning of *Faust, Part Two*. After an indefinite length of time following Gretchen's death in the tragic ending of *Part One*, Faust awakens at sunrise restored by nature. Nature is personified by Ariel and a chorus of spirits. Faust speaks of how his life is rejuvenated by their healing powers. His life pulse beats anew, and addressing the earth he exclaims, "Du regst und rührst ein kräftiges Beschließen, / Zum höchsten Dasein immerfort zu streben" ("You promote and astir a strong resolution to strive unceasingly to highest being" [3: 148, 4684–5]). Although Goethe could be referring to Faust's own reinvigorated striving for the highest peaks of existence, he is equally describing the active determination within nature herself. It is this intent or force that the homeopathic remedy claims to embody. The dynamism in the remedy resembles the reviving miniature spirits of nature that accompany Ariel. By lending voice to Ariel and his elves, Goethe poetically represents and acknowledges the magic-like effects of nature, alone capable of restoring Faust.

Writing in a different context on the history of *Lebenskraft*, Maike Arz notes that "the vitalistic model of nature implies that the human being is called through a natural power that resides within him, on the one hand, to self-realization [*Selbstverwirklichung*] and, on the other, to participation in the conjoined fate of all living beings" (35). Vitalism, in her words, entails both the self-determination of man and a vocation to see human life in comparison or analogy with the natural world. Precisely this dual context inhabits the phrase "ein kräftiges Beschließen": Goethe conjoins the activity of nature and Faust's own determination.

Novalis, too, visualized parallels between the forces in nature and those that govern the human being, in particular, the human body. Understanding the human body, for him, was the prerequisite to peering into the heart of things:

> The essence of what stimulates us is called nature, and therefore nature stands in an immediate relation to those functions of our bodies that we

call senses. Unknown and marvellous relations of our body allow us to imagine unknown and marvellous correlations with nature. Hence, nature is that wonderful community into which our bodies introduce us, and which we learn to know in the measure of our body's faculties and abilities ... One can see that these inner relations and faculties of our body must be studied before all else, before we ... unravel the nature of things. (1: 220–1).[2]

Hahnemann likewise believed that the human organism reflected the essence of nature. In deriving the principle of *similia similibus curentur* from the reaction of the human body, he advanced insight into these "unknown and mysterious relationships." Healing was to be the proof that invisible powers, traversing the natural world to mankind, actually existed.

Schelling offers another prime example of how writers of the period established parallels between humanity and a vital spirit-like force in the vegetable world. "Every plant," he wrote in *System des transcendentalen Idealismus (System of Transcendental Idealism)*, "is a symbol of intelligence" (*Werke* 2: 490). Indeed, the morphological structure or organization of a plant acted symbolically: it embodied a productive force, the force of spirit. Each plant, so to speak, traced the intricate pattern of the soul.

> Therefore there is something inherently *symbolic* about each organization, and each plant is, as a manner of speaking, *an arabesque delineation of the soul* [*der verschlungene Zug der Seele*]. Since our spirit possesses the infinite desire to organize ourselves, the outer world must reveal a general tendency to organization ... There is a productive force in matter around us. Such a force is but only the force of a *spirit*. (*Historisch-Kritische Ausgabe* 1.4: 113–14)

In Schellingian terms, we are as far from Kant's dualism as from Fichte's focus on the self. The subject is not opposed to nature, because the subject is part of nature. Homeopathy would not work without the analogous premise that there is an organic force pervading all nature, of which the human is part. In this biomorphic universe, herb-based remedies have vital powers, just as humans do. Indeed, according to the *"fixed and eternal laws of nature"* (*Organon* 157, §111), each plant, mineral, and salt displays its own peculiar property (*Organon* 157, 159–60, §§ 111, 118–19), especially on the psychic state of a patient (*Organon*

192, §212). As Goethe famously wrote: "Jede Planze verkündet dir nun die ew'gen Gesetze, / Jede Blume, sie spricht lauter und lauter mit dir" ("Each plant heralds now to you the eternal laws, each flower addresses you louder and even louder"; HA 13: 109).

Here the Romantic longing for absolute unity, for realizing infinite spirit within the finite world, joins hands with Hahnemann's declaration that homeopathy is grounded in empiricism and experience. To put it differently, one cannot excise the metaphysical dimensions out of homeopathy to claim that it is based purely on observational proof. Even Hahnemann's contemporaneous critics said as much. The Hanoverian court physician Johann Stieglitz pronounced that homeopathy belonged to metaphysics and speculative *Naturphilosophie*. He derided it for claiming to create spirit out of purely mechanical means – rubbing and shaking (124). Another prominent physician of the period – and among the first to advocate the smallpox vaccination – Georg Christian Gottlieb Wedekind (1761–1831), pointed out that Hahnemann misapplied metaphysical categories to living organisms (24), just as the *Naturphilosophen* did (25). In his 1799 *Erster Entwurf eines Systems der Naturphilosophie* (*First Outline of a System of the Philosophy of Nature*) Schelling, applying John Brown's concept of *Erregbarkeit* to the "*dynamic forces* of the universe (*Werke* 2: 9), claimed that "to philosophize nature is to *create* nature" (*Werke* 2: 13). Hahnemann, of course, was critical of the pure speculation of the *Naturphilosophen*. But homeopathy creates nature much as Schelling's philosophy does: not only does it envisage beneficent unifying forces within nature, it claims to harness them. Hahnemann would have agreed with Schiller's assessment that "philosophy and the science of medicine stand beside each other in perfect harmony" (20: 38), and that his own medical practice best represented this congruence.[3] Haehl quotes him as saying: "Philosophy represents the highest ideals towards which the human mind is imbued with the desire to struggle. Philosophy is not only the highest of all sciences, it is also the basis and the fundamentals of all others. No science can exist without philosophy, for without its help it falls to the level of the handicraft . . . This is true above all of medicine" (251). As mentioned in the introduction, none other than the famous novelist of the period, John Paul Richter, characterized Hahnemann as an "an odd Janus head [Doppelkopf] of philosophy and studiousness" (292).

The father of homeopathy thus would have seconded Novalis that the true observer of nature needed to be as ingenious and creative as she is. Out of manifold external phenomena, the true empiricist needed to distil the essential. "To experimenting belongs *natural genius*, that is, that wondrous ability to capture the sense of nature – and to act in her spirit. The true observer is an *artist* – he *divines* the *significant* and knows how to ascertain what is crucial out of the strange, fleeting mixture of the appearances" (2: 471). This true observation is precisely what Hahnemann claimed to accomplish. Out of varied empirical phenomena he purported to extract therapeutic principles: he grasped despite myriad symptoms of infirmity a benevolent, unifying wellspring of healing. Another luminary of the period, Alexander von Humboldt, wrote to Schelling in 1805 after his travels in the Americas: "*Naturphilosophie* cannot harm the progress of the empirical sciences. On the contrary, it traces a discovery back to its principles and simultaneously provides the foundation for new discoveries" (Schelling, *Briefe und Dokumente*, 3: 181). In the assessment of Germany's leading historian of Romantic science, Dietrich von Engelhardt, "Romantic natural science did not want in any way to oppose empiricism; rather, physics and metaphysics were to be combined. The empirical standpoint was to be supplemented, not abolished" ("Natural Science in the Age of Romanticism" 123).

The reason underlying this inseparability of empiricism and metaphysics,[4] of the physical and the spiritual, is that nature, in and despite her manifest, observable particulars, is for the Romantics always one, whole, and spirit-infused. They exuberantly affirmed the Spinozistic oneness of God and nature and hence that spirit enlivened all of nature, from the inanimate to the living, and the smallest particle to the largest being. Be it Herder, Goethe, Hölderlin, or Novalis, they all fervently believed in the Spinozistic concept of an infinitely extensive nature identical with God – of *hen kai pan* (or *Alleinheit*). In the words of Charles Taylor, "Herder and those of his generation and the succeeding one were . . . greatly influenced by Spinoza . . . What Spinoza seemed to offer, why he drew Goethe, and tempted so many others, was a vision of the way in which the finite subject fitted into a universal current of life. In the process Spinoza was pushed towards a kind of pantheism of a universal life force" (*Hegel* 16).[5] We know from Richard Haehl that, from his school days onward, Hahnemann had followed Spinoza (as well as Leibniz) (251).

In Gilles Deleuze's phrasing, according to Spinoza there is "one Nature for all bodies, one Nature for all individuals, a Nature that is itself an individual varying in an infinite number of ways" (*Spinoza* 122). In Novalis's words, "the life of the universe [is] a hundred-voiced conversation" (1: 230). In 1784, after mentioning to Karl Ludwig von Knebel that he was reading Spinoza's *Ethics*, Goethe enthused that every creature was "only a tone, a shade of one great harmony" (*Briefwechsel* 55). Schelling was to take this belief further, maintaining that both the inorganic and organic worlds, in other words, all material phenomena, were manifestations of the Absolute.[6] Dietrich von Engelhardt concludes: "According to the romantic view, the multiplicity of natural phenomena and the difference between inorganic and organic nature cannot conceal the connection and the unity of nature" ("Natural Science in the Age of Romanticism" 109). This Romantic tradition stretched beyond Germany: Coleridge, in his 1816 essay "Theory of Life," referred to the indivisibility of life as "the existence of all in each as a condition of Nature's unity and substantiality, and of the latency under the predominance of some one power, as wherein subsists her life and its endless variety" (525). He closes by emphasizing: "*Thus*, then, Life itself is not a *Thing* – a self-subsistent *Hypostasis* – but an *Act* and *Process*" (557). By choosing the words "*power*," "*act*," and "*process*," Coleridge clearly highlights the dynamism inherent in life.

In an unpublished essay, "Spinoza's God in Goethe's Leaf," arguing that Goethe's "entire scientific vision was clearly articulated in Spinozist terms," Michail Vlasopoulos explains that "the version of extension that Spinozist physics takes as its object is *inherently and eternally dynamic*, unlike Descartes." In the previous chapter we saw how Hahnemann repeatedly characterized the action of the homeopathic remedy as kinetic as well as spirit-like. Translated into Spinozist vocabulary, the plant-based remedy can be seen to *express* infinite vibrancy, each plant doing so in a certain and determinant mode unique to it. Gilles Deleuze outlines these two steps at the start of his tome *Expressionism in Philosophy: Spinoza*, citing the first part, definition six of Spinoza's *Ethics*: "By God I understand a being absolutely infinite, that is, a substance consisting of an infinity of attributes, of which each one *expresses* an eternal and infinite essence" (13). And then, still quoting Spinoza: "Whatever exists expresses the nature or essence of God in a certain and determinant way" (14). A Spinozist framework helps to justify how Hahnemann could postulate (1) a supersensible healing action (a kind of *natura naturans*) as (2) empirically concretized, that is, directly manifested in the natural remedy (*natura naturata*),[7] in fact, even expressed further via trituration and shaking. The remedy is an extension or emanation of this one living unity of nature. The supersensible is thus at one and the same time empirically real.[8]

Alexander von Humboldt, Oken, Ritter, G.H. Schubert, and Steffens were all natural scientists of the period who believed in this vibrant interconnectedness of nature, whose unity it was their calling to explore. Ritter wrote: "Anyone who finds in infinite nature nothing but one whole, one complete poem, in whose every word, every syllable, the harmony of the whole rings out and nothing destroys it, has won the highest prize of all" (*Fragmente* 2: 205). Schubert likewise enthused: "The history of nature has to do not just with individual, finite, imminently perishable being, but with an imperishable basis of all that can be seen, which unites it all and gives it soul" (*Allgemeine Naturgeschichte* 4). The Danish scientist and philosopher Steffens echoed this feeling as well: "Whomever nature allows to discover her harmony within himself – bears a whole, infinite world in his inner being – he is the most individual creation – and the most sacred priest of nature" (317). Knowledge of nature was a branch of the self. Thus, in his *Kosmos*, Alexander von Humboldt wrote that his aim as a scientist was "to arrive at a higher point of view, from which all formations and forces reveal themselves as one, living, internally active whole of nature. Nature is not a dead aggregate. She is "for the enthusiastic researcher," as Schelling expressed it in his wonderful essay on the plastic arts, "the holy, eternally creative, primary force of the world, who actively generates and produces all things out of herself" (1: 39). Similar passages can be found throughout the writings of the period.

There are three important ways in which the homeopathic cure performs this symbiosis between man and an all-enlivening nature. First, the efficacy of nature, her inherent oneness, allows for the homeopathic system of analogy between symptoms produced by disease and those produced by the remedy. At the base of homeopathy is a theory of interconnectedness, even to the point of disease antidoting disease. Homeopathy thereby resolves any antagonism between world and self. In its world view, moreover, nature does not allow anything to be hidden. Plants in their biodiversity provide all that is necessary to restore the human body to health and in a manner that transforms toxicity into an instrument of prompt, mild, and permanent healing (*Organon* 95, §2).

Second, despite the bewildering proliferation of discrepant pathological indexes that Hahnemann records in his patient journals, protocol books, repertoires, and *materia medica*, he can be reassured that nature in her generosity will provide for healing in accordance with her laws, as revealed by homeopathy. The physician need not be perturbed by this proliferation, for the oneness of nature, especially as testified by the all-pervasive vital force, overcomes differences. Indeed, as Novalis put it, the strangeness of the particular lends testimony to the greatness of the whole: "The *greater* and more complex the whole, the more remarkable the particular" (2: 645, #717). In truth, with his dynamic view of illness, Hahnemann believed that every symptom, every single part of the body, affected the whole and reflected the imbalance of *Lebenskraft*.

Third, and most important, with regard to the Law of Minimum, whereby a substance is not just still present but in fact activated after exponential dilution, "every infinite is actual" (Deleuze, *Spinoza* 79) and the infinite is present in the part. According to Spinoza, everything that exists is a part and expression of God, that is, Nature and Substance. Hahnemann, as it were, literalizes Spinoza: active substance is expressed wholly in the minuscule part. He could thus enlist the minute and minimal against the plethora of illnesses. The infinitesimal dose refines and clarifies the latent energy of nature. In other words, it is not the case that a minuscule, portioned amount of the original toxin accounts for the remedy's effectiveness. Spinoza stated: "An absolutely infinite substance is indivisible" (85, 1P13). Goethe in glossing Spinoza remarks: "The infinite cannot be said to have parts" (13: 7).[9] The effectiveness of the homeopathic solution, accordingly, comes not from the mathematical reduction or from a material residue but from the infinite presence of nature that is allowed to unfold and emanate in the dynamized remedy. The Law of Minimum illustrates the presence of an infinite action in the finite, minuscule dose. The homeopath would thereby realize what Schelling envisaged as the desideratum of all sciences: to represent the infinite within the finite ("Möglichkeit der Darstellung des Unendlichen im Endlichen – ist höchstes Problem aller Wissenschaften" [*Werke* 2: 14]).

Various authors of the period used the natural metaphors of emanation, waves, galvanic fluids, and resonances to illustrate this interconnectedness of an infinite nature. Their world is not conceived as unconnected particulars. Goethe surmised in his 1792 essay "Der Versuch als Vermittler von Objekt und Subjekt" ("The Experiment as Mediator between Object and Subject"): "Nothing happens in living nature that does not bear some relation to the whole . . . All things in nature, especially the commoner forces and elements work incessantly upon one another; we can say that each phenomenon is connected with countless others just as we can say that a point of light floating in space sends its rays in all directions" (13: 17–18).[10] Coleridge similarly gives as an example of the "essential vitality of Nature" that she "expands as the concentric circles on the lake from the point to which the stone in its fall had given the first impulse" (509). He also cites "the arborescent forms on a frosty morning, to be seen on the window and pavement" as having "*some* relation to the more perfect forms developed in the vegetable world" (508). Nature's vitality operates in both the organic and inorganic realms and is visibly evident in such patterning between the two.

Novalis also was fascinated by the image of frost, snow, and crystals – their intricate patterns and atomic transmutation. Resemblances between the designs on eggshells, bird's wings, clouds, and plants illustrated how nature hosted diverse similarities that promised to manifest the oneness unifying them. At the start of *Die Lehrlinge zu Sais* he declared that a delicate writing can be detected everywhere in such objects – a script of ciphers that promises to reveal a magical inner correspondence operative within nature. They were full of life and evidence of the eternally pulsing organisms of the earth. Novalis admits reluctantly, however, that the key to unlock their language cannot be found (1: 201). They resembled "strange conjectures of chance" (ibid.). He concedes elsewhere that "the meaning of the hieroglyph is missing" (2: 334, #104) and that the goal of Romanticism was to reverse this state of affairs: "The world must be romanticized. It is the way to recover original meaning" (2: 234, #105).

Had Hahnemann, though, found the grammar or poetic system of analogy that decoded the book of nature that the Romantics had sought? Could it be that he had discovered the few magical principles that solved the mystery of diseases and their cure via nature's munificence? Did he render the inaccessible at hand through analogical thinking? Did the homeopathic cure assist "the living organism" to react never "beyond what is absolutely necessary" back "to the natural state of health" (*Organon* 157, §112)? However anyone today would respond to these questions is not the issue: Hahnemann himself would have confidently answered in the affirmative for the intellectual climate in which he lived allowed him to do so. His system is as universalizing as those of the German Idealists from Kant to Hegel. Like Schelling, in particular, he absorbs spirit into nature and nature into spirit. He rethinks the life force as material. But he is unlike these philosophers in his actual testing of substances. He does not remain on the level of theory. Unlike them, he felt that *Leben* could be parsed semiotically, and he set out in his praxis to develop technical command through the studying of detail.

To put it differently, it is essential to be able to account for this disparity between, on the one hand, attention to the empirical detailed observation in Hahnemann's praxis and, on the other, how he searches for principles of knowledge, an idea, or whole vision grounded in nature that will unify the varying parts. It is a division that explains how, with one foot, the father of homeopathy stands in the Enlightenment and, with the other, becomes a Romantic. Semiotics provided the basis for comparing symptoms, but the basis of healing is reliant on the dynamic action of the remedy. Hahnemann needed to move increasingly, as his career progressed, to a concept of a *force* that would establish relationality between human and natural worlds.

The popularity of Spinoza's *hen kai pan* at the time helps in large part to explain how Hahnemann could reconcile the part and whole, the visible and invisible; that he is both empirical and speculative. But there is another unlocking key here, namely, Goethe's notion of *anschauende Urteilskraft* (intuitive power of judgment). In writing on the influence of Spinoza on Goethe, Frederick Amrine characterizes Spinoza's *scientia intuitiva* as "a non-discursive, synoptic perception of nature in its entirety, of thinking of wholeness in its immediacy . . . the kind of practice Goethe exemplifies in his morphological studies" ("Goethean Intuitions" 40). In 1792 the very unity of all the particulars in nature raised for Goethe "just this question: how do we find the link that holds these phenomena together[?]" (HA 13: 17). The succinct answer appeared twenty-eight years later in the essay "Anschauende Urteilskraft": "Through the intuitive study [Anschauen] of a constantly creating nature we render ourselves worthy of intellectually participating in her productions" (13: 30–1).

For Goethe, *anschauende Urteilskraft* permitted both the artist and the scientist to connect the universal and particular. This holistic, synthetic mode of thinking established links between man and nature, and between things in nature. Moreover, such immediate "seeing" merges practical experience with insight into the totality of things. It arose when Goethe set about to study the grammar of botany in the earliest version of *Die Metamorphose der Pflanzen* (*The Metamorphosis of Plants*) of 1790. But even as early as his *Urfaust*, Goethe formulated this desire to comprehend the unity of things: "Daß ich erkenne, was die Welt / Im Innersten zusammenhält, / Schau alle Würkungskraft und Samen" ("That I may discern whatever / Binds the world's innermost core together, / See all its active forces, and its seeds" [HA 3: 367, 29–31]).

Conclusion

Reflecting on his decades of studying plants later in 1820, Goethe acknowledged vegetal "origination from a lively mysterious whole"; nature had "brought together states seemingly foreign to one another and united them as a whole" (HA 13: 27). Noting how plant filaments flowed together (anastomosis), Goethe believed an inner law residing within the plant governed its organic growth and manifested itself in various forms. Despite their diversity, the separate vegetal organs are one and the same: from the individual attributes of a plant one can perceive its unifying wholeness. Goethe writes: "Whether the plant sprouts, blooms, or carries fruit, it is always *the same organs*, which, despite manifold purposes and in the guise of often variegated forms, fulfill the prescription of nature" (HA 13: 100, §115). In other words, by paying close attention to the disparate parts of a plant the observer can intuit one generative unity behind them. Moreover, this desire to grasp the totality of a given phenomenon as a unified totality via *anschauende Urteilskraft* led Goethe to "see" how an archetypal plant (*Urpflanze* and *Urphänomen*) manifested itself in various plant forms. In short, the distinctive phenomena in nature are linked to the entirety of nature and reveal its inner, unseen, teleological forces. Goethe believed one could come to this essential form only through close observation of nature; the prerequisite for perceiving analogies intuitively was an empirically trained eye. It was thus that the unifying form seen in-between its individual manifestations could be very much real for Goethe.

This intuitive act of immediately synthesizing, grasping the fundamental way parts fit into a whole, thus making visible the invisible, is something that Kant refused to accept. John Smith and Elizabeth Millán in their introduction to a special issue of the *Goethe Yearbook*, "Goethe and Idealism," wrote that paragraphs 76 and 77 of the *Critique of Judgment*

> piqued Goethe's interest and opened avenues that moved beyond Kant himself. [It was here] that Kant raised the possibility of a "intuitive intellect," an *intellectus archetypus*, that could grasp the kind of teleological unity-in-diversity that makes living organisms unique and might even provide a model for all of nature as itself a living organism. Whereas Kant denied this faculty to humans, whose reason could only proceed discursively, such intuitive knowledge was at the heart of not only Goethe's poetics and scientific thought, but also . . . at the heart of his very sense of self. (6)

And Goethe was not alone: the Romantics in turn were to speak of a Spinozistic *intellektuelle Anschauung*,[11] much in the same way as they characterized *Potenzierung*. Novalis famously portrayed Romanticism as the clairvoyant activity of elevation (*Potenzierung*) that endows the commonplace with a higher meaning (2: 334, #105).

Like Goethe, who sought "the inner identity of different plant parts" (HA 13: 82, §60), so too Hahnemann searched beyond the mere attributes or effects of a remedy towards the essence of its spirit-like action, which was a manifestation of one enlivening protean force residing within nature. He wanted to capture the immaterial powers residing in a plant, for, were it not for *Potenzierung*, the active ingredient in a fresh herb would decay and rot (*Organon* 215, §266). "Im innern Wesen" (*Organon-Synopse* 281, §20) and "inwohnende" (*Organon-Synopse* 525, §117) were words (that may be translated as "innate") that he frequently used to characterize this hidden power. Trituration and shaking would draw forth and exalt the power that lay dormant and concealed in the innermost being of the plant (*Organon* 162–3, §128). Like Faust, then, Hahnemann wanted to reveal "was die Welt / *Im Innersten* zusammenhält" (italics mine [HA 3: 367, 29–30]). One may recall that the 1790s likewise saw Goethe develop his theory of the symbol, which is to say, the making visible of the invisible. The infinitesimal homeopathic quantity holds, like the symbol, the essence of what it represents: it exemplifies, even in miniature, the living, active presence in nature. Indeed, more than just signifying this unseen presence, thanks to *Potenzierung*, it actually bodies it forth. Similarly securing a vocabulary that would express the potential for transformation, Goethe spoke of expansion and contraction, diastole and systole, inhaling and exhaling (HA 12: 436; 13: 337 and 488). The poet also adopted the language of infinitization: in writing to Herder on 17 May 1787, he avowed the unifying concept of the *Urpflanze* to be the model and key to creating plants into infinity ("Pflanzen ins Unendliche erfinden" HA 11: 324).

With this Goethean framework in mind, Hahnemann's reasoning in conjunction with the term *Erfahrung* in the *Organon der Heilkunst* takes on new meaning. Section 28 of the *Organon* reads:

> Da dieses Naturheilgesetz sich in allen reinen Versuchen und allen ächten Erfahrungen der Welt beurkundet, die Tatsache also besteht, so kommt auf die scientifische Erklärung, wie dieß zugehe, wenig an und ich setze wenig Werth darauf, dergleichen zu versuchen. Doch bewährt sich

folgende Ansicht als die wahrscheinlichste, da sie sich auf lauter Erfahrungs-Prämissen gründet. (*Organon-Synopse* 297)

> As this therapeutic law of nature clearly manifests itself in every accurate experiment and research, it consequently becomes an established fact, however unsatisfactory may be the scientific theory of the manner in which it takes place. I attach no value whatever to any explanation that could be given on this head; yet the following view of the subject appears to me to be the most reasonable, because it is founded upon experimental premises. (107)

Hahnemann appeals to nature to ground the laws of homeopathy: its laws, confirmed "in allen reinen Versuchen und allen ächten Erfahrungen der Welt," become his own. Consequently, in §26 he speaks of the homeopathic *Naturgesetz* (*Organon-Synopse* 291). When he warns of "all conjecture, fiction, or gratuitous assertion," it is in opposition to "nothing but the pure language of nature, carefully and honestly interrogated" (*Organon* 169, §144). This, then, is for him true *Erfahrung*: it is set in opposition to an older school of medicine that would rely on the microscope and hypothetical designations of disease (*Organon* 105, §25), in other words, a language which is not nature's own. Hahnemann, by contrast, as the opening sentence indicates, confirms ("besteht") the mysterious workings of nature through *Erfahrung*. But these forces nonetheless remain mysterious and need not be explained ("so kommt auf die scientifische Erklärung, *wie dieß zugehe*, wenig an und ich setze wenig Werth darauf").

To be more precise, at the close of this passage on *Naturheilgesetz* and *Erfahrung*, Hahnemann does not say that homeopathy is based on *Erfahrung* itself, but on premises derived from a sense of probability, that which is "wahrscheinlichst." He then refers to *Erfahrungs-Prämissen*: empirical observations, reliant on the authenticity of personal experience, form the basis of his premises. The laws or principles of homeopathy are built on these premises. To put it another way, Hahnemann turns conjecture into observable law, in fact, *Naturheilgesetz*, and professes to fill in evidential gaps. However difficult it may be for us to follow this line of reasoning today, it is a thought process that is emblematic for its time. In Goethean terms, the homeopath immediately intuits the whole – the innermost essence and unity of things – from the observable parts. Already in 1792 Goethe conceptualized the task of the scientific researcher as working towards such a coalescing vision, towards empirical evidence of this higher sort ("Erfahrungen der höheren Art" [HA 13: 18]).[12] *Scientia* is here conceived as actual, unifying, and experiential knowledge, not just a collection of facts.[13]

In sum, the father of homeopathy pursued and formulated principles of knowledge that would unify the disparate, incongruent observations in his praxis. Like many post-Kantians of his time, he believed that a dynamic force internally organized nature and man in their innermost being and that scientists could provide evidence for this force. But just as the botanist Goethe developed Kant into realms the philosopher cautioned against, so too did Hahnemann. To be sure, Hahnemann alluded to Kant: "The unprejudiced observer (however great may be his powers of penetration) is aware of the futility of all elaborate speculations that are not confirmed by experience [*Erfahrung*]" (*Organon* 96, §6).[14] Set in Kantian terms, however, the laws of homeopathy are not just regulative ideas, they are constituent of nature. They are not just hypothetical constructs but form a systematic science. However much Hahnemann claimed at the time that his laws were empirically grounded, Kant would have rejected them. The post-Kantians, in which we can include Hahnemann, would not have. They, unlike Kant, posited a dynamic force in nature and were firm in their belief that it could be proved.

The Birth of Homeopathy out of the Spirit of Romanticism has endeavoured to investigate this intellectual world that surrounded Hahnemann's writings. Other authors around 1800 contextualize and lend meaning to Hahnemann's pristinely laid out system whose structural wholeness was intended to mirror the unity between spirit and nature. But I want to go one step further and suggest not only that homeopathy was a product of its time; I also want to submit that only by virtue of this intellectual framework did homeopathy enjoy the widespread resonance it did. By adopting culturally relevant paradigms, Hahnemann ensured an attentive audience. One notices how Hahnemann incorporated into his vocabulary as his career progressed terms that had garnered currency, such as *Lebenskraft* and *Potenzierung*. In repeatedly referring to homeopathy as nature's law, he invoked the Romantic concept of harmony between man and nature. In addition, he turned to the sensational healing modality of mesmerism and even characterized it as a form of homeopathy. With his attentive bedside manner and his encouragement of patients to self-monitor their symptoms, diet, and progress, Hahnemann appealed to a growing middle-class self-consciousness and sense of unique individuality; the lengthy medical consultation met the patient's desire for open, transparent, comprehensive communication. Homeopathy thereby transfigured the commonplace of illness.

Into the bargain, Hahnemann's self-characterization as innovator and authority falls in line with the cult of genius at the time. As we have seen, his reliance on self-testing echoes the notion common among scientists at the time that self-experimentation guaranteed authenticity and truth. As well, only the insightful, intuitive intellect could perceive among myriad pathological symptoms the unusual single indicator that provided the key to the choice of remedy. And only the brilliant scientist could discern the dynamic, spirit-like action of these remedies, draw them out through a process of potentization, and thereby impart the gift of natural healing to his patients.

Notes

Introduction

1 Macrobiotics was a term coined by Hufeland and today, as in Hufeland's work, makes recommendations for influencing one's health or what he called the vital life force (*Lebenskraft*) via diet and lifestyle.
2 Citations from the *Hamburger Ausgabe* of Goethe's works are abbreviated as HA. Citations from the *Weimarer Ausgabe* are abbreviated as WA. Citations from the *Gedenkausgabe* are abbreviated as GA. Unless otherwise noted, I take responsibility for all translations.
3 On Bettina von Arnim's engagement for homeopathy, see Dinges ("Bettine von Arnim"); and Schiffter.
4 If we look at the major histories of medicine, Roy Porter (*Greatest Benefit*) devotes a few pages to homeopathy under the category "Alternative Medicine," while Karl Rothschuh isolates homeopathy under the category "Biodynamic Medical Concepts" (*Konzepte der Medizin*). For general histories of homeopathy, see Fink; Heinze; Seiler; Tischner (*Geschichte der Homöopathie*); and Wischner (*Kleine Geschichte*). On the history of homeopathy in America, see John H. Haller.
5 A note on the translations of Hahnemann: I have chosen to cite from the existing translations of Hahnemann's works, despite their age and infelicities, because of how widespread their usage is. When referring to Hahnemann's major work, the *Organon der Heilkunst*, I cite the translated edition by Constantine Hering of 1849, entitled *Organon of Homeopathic Medicine*, and I also refer to specific paragraphs within it (designated by the sign §). When citing earlier editions, the translations are my own. When it is necessary for better comprehension, I indicate the specific German term.

6 The founder of homeopathy did not circulate among the Romantics who were in constant dialogue among themselves. Apart from his adoption of mesmerism and one reference to infinitesimal calculus, Hahnemann also did not orient his findings in terms of other sciences. Wischner, in fact, observes that, after 1822, Hahnemann did not even read scientific journals any more (*Fortschritt oder Sackgasse?* 16). One therefore needs to be careful not to leave the impression that Hahnemann was in direct communication with the major thinkers of his era. Yet so too one has to be careful not to remove homeopathy from the era of its creation, lest we forget that, like Romantic literature, philosophy, and life sciences, homeopathy is steeped in concepts of analogy, *Lebenskraft*, and *Potenzierung*.

7 To my mind, the most balanced overview of the intellectual and conceptual history informing homeopathy, as well as a summary of its reception history, is Schmidt, "Entstehung, Verbreitung und Entwicklung." Schmidt argues that one needs to understand Hahnemann in his historical context to see how he reasoned and how he was debated. He here also summarizes tension within Hahnemann's thought (68). Schmidt has further contributed to the historical positioning of Hahnemann through *Die philosophischen Vorstellungen Samuel Hahnemanns bei der Begründung der Homöopathie*, where he presents Hahnemann exclusively as an Enlightenment thinker. See also Fräntzki, *Die Idee der Wissenschaft bei Samuel Hahnemann*.

8 Still, scholarship on Romantic medicine and science only incidentally references homeopathy. One searches in vain for anything more than a passing entry on Hahnemann (Aesch; Borgards; Botsch; Brandstetter et al; Cunningham and Jardine; Engelhardt; Gerabek; Gigante; Holland; Holmes; Leibbrand; Lenoir; Lohff; Mischer; Mocek; Neubauer; Reill; Richards; Steigerwald; Tsouyopoulos (*Asklepios*); Wallen; Wetzels; Wiesing). Only a couple of books on medicine in the age of Goethe briefly mention him (Buchinger; Pfeiffer; Tobin). Others do not at all (Egger; Zumbusch).

9 See also Risse, "'Philosophical' Medicine in Nineteenth-Century Germany."

10 The established biography is by Haehl. See also Gawlik; Handley. A recent comprehensive biography, which has the advantage of being online, is Jütte (*Samuel Hahnemann: The Founder of Homeopathy*).

11 See Hahnemann's essay "Striche zur Schilderung Klockenbrings während seines Trübsinns" ("Description of Klockenbring during his Insanity" [1796]).

12 As noted by Schmidt in "Samuel Hahnemann und das Ähnlichkeitsprinzip" (161–2).
13 In the essay "Fingerzeige auf den homöopathischen Gebrauch der Arzneien in der bisherigen Praxis" (*Gesammelte kleine Schriften* 461).
14 See §§ 2, 19, 33, 45, 126, and 200 of the *Organon*.
15 A comparison of the various versions has been facilitated by *Organon-Synopse: Die 6 Auflagen von 1810–1842 im Überblick*. It reprints the editions parallel to one another. For a critical edition of the 6th *Organon*, with systematic commentary and a glossary, see Schmidt's edition. For a similarly indispensable glossary of all of Hahnemann's work, see Minder.
16 For the sake of philological comparison: at about the same time (1811), the editors of the *Allgemeine medizinische Annalen* separated their monthly journal into two separate issues, "an *Annalen der Heilkunde* containing materials of theoretical interest, and an *Annalen der Heilkunst* aimed at practitioners" (Broman 159). See also Schmidt's edition of the *Organon* (346–7) for a summary of how Hahnemann used the two terms in this work.
17 On the history of the reception of the *Organon* worldwide, see Baur.
18 On the *Fragmenta* see the edition and analysis by Wettemann. For a parallel study on the case journals from 1803–6, see *Die Pharmakotherapie Samuel Hahnemanns in der Frühzeit der Homöopathie: Edition und Kommentar des Krankenjournals Nr. 5* by Varady.
19 The *Gesammelte Arzneimittellehre* edited by Lucae and Wischner brings together all of Hahnemann's remedies in one volume.
20 The view of Hahnemann as an Enlightenment thinker has dominated scholarship on him. In particular see Brockmeyer (chapter on "Gott und Arzt"; Große-Onnnebrink; and Schmidt (*Die philosophischen Vorstellungen*). In his edition to the *Organon* Schmidt has also collected Hahnemann's deistic references to the goodness and wisdom of the Creator (290, 342–3).
21 Ziolkowski refers to the year between Spring 1794 and Summer 1795 as the *annus mirabilis jenensis* – das Wunderjahr in Jena.
22 Schmidt makes the interesting observation that the afterlife of homeopathy is due to the paradoxical reason that it cannot be proved true or false ("Die Entstehung, Verbreitung und Entwicklung" 70).
23 To repeat, there are three sorts of collections of symptoms: the *materia medica*, Hahnemann's repository, and the patient records. In addition, there is a fourth, one that, like the repository, has not been transcribed and edited. These are the very few extant notebooks of Hahnemann where he jotted down his provings – the testing of a substance on himself. In actuality, these extant notebooks record not just Hahnemann's self-testings

or those on his students but numerous excerpts of symptoms copied from *materia medica* not his own, such as Cullen's. Some of these excerpts have found their way into the published *Reine Arzneimittellehre*, an issue taken up at the close of chapter 2.

24 For access to the library's holdings as well as detailed information on the IGM and their work including online publications, see http://www.igm-bosch.de/content/language1/html/index.asp.
25 On this distinction see Dinges, "Zum Standard der Forschung" 25.
26 In particular, see the collections edited by Martin Dinges, *Homöopathie: Patienten, Heilkundige, Institutionen* and *Patients in the History of Homeopathy*.
27 Another scholar who approaches the history of German medicine as an *Alltagsgeschichte* is Lindemann. See also the collections on the social history of medicine by Labisch and Spree; Lachmund and Stollberg. Other topics they address include the social history of confinement, specific practices such as abortion or contraception, and the social history of communicative diseases. For an overview of the different approaches to the history of medicine, consult Eckart and Jütte, eds; Norbert and Schlich, eds.

Chapter One

1 For investigations of *similia similibus curentur*, see Just; Jütte ("200 Jahre Simile-Prinzip: Magie–Medizin–Metaphor"); Carl Werner Müller; and Schmidt ("Samuel Hahnemann und das Ähnlichkeitsprinzip").
2 He similarly writes: "It has always been a matter worthy of the greatest admiration to see how nature, without having recourse to any surgical operation, without having access to any remedy from without, does often when left quite unassisted, develop from itself invisible operations whereby it is able . . . to remove diseases and affections of many kinds. But she does not do these for our imitation!" ("The Medicine of Experience," *Lesser Writings* 439).
3 Other medicinal substances he treats in the 1796 essay include *strychnos nux vomica, digitalia purpurea, datura stramonium, nicotiana tabacum, atropa belladonna, oethusa cynapium, arnica montana, ignatia amara,* and *ipecacuanha*.
4 For the main study on Hahnemann's self-experiment with Peruvian bark, see Bayr.
5 See Wiesemann.
6 For a review of the history of opium, see Maehle.
7 See also Broman (144) on opium and its enthusiastic reception by German Brunonians.

8 For a study of the different stages in development of the theory *similia similibus curentur* from 1796 onward, see Schmidt, "Samuel Hahnemann und das Ähnlichkeitsprinzip," in particular the helpful summary (167). Schmidt argues that between 1805 and 1810 Hahnemann is trying to ratify, rationalize, and theorize his vision, but that this speculative theory – by which Schmidt means not only the early working hypothesis of primary and secondary effects but also the post-1805 postulates such as the unity of the organism, the incompatibility of two simultaneous irritations in the body, the identity of disease with its symptoms, and the irrelevancy of the cause of diseases (174) – does not contraindicate the effectiveness of homeopathic remedies and praxis.

9 Hahnemann also noted, in the case of scarlet fever, that the progression of an illness was biphasic ("Cure and Prevention of Scarlet-fever," *Lesser Writings* 374). And he observed the after-sufferings once the disease was over (377).

10 Hahnemann's belief that two competing illnesses could not reside simultaneously in the body was not unusual in the eighteenth century. Roy Porter refers to "the old saw that diseases were jealous of each other and mutually exclusive. So long as gout was in possession, no deadlier enemy could gain invasion" ("What Is Disease?" 93).

11 Schmidt ("Samuel Hahnemann und das Ähnlichkeitsprinzip" 165) notes that it was around 1805 that Hahnemann began to use such comparisons in order to support his claim for the general applicability of *similia similibus curentur*. For more on the references Hahnemann cites to bolster his Law of Similars, see Schmidt, "Die literarischen Belege Samuel Hahnemanns für das Simile-Prinzip."

12 See the article by Watzke.

13 Volker Hess also situates Hahnemann in terms of eighteenth-century semiotics, though his focus is on the transparency of the sign. He writes: "This passage is an illustration of the peculiar transparency typical for 18th-century thought. The problem of coding did not exist. The signifier did not obscure the signified" ("Zeichen ohne Differenz" 75). I would insist, by contrast, that in homeopathic semiotics signs do not refer to a signified but to other signs.

14 Only a few of the protocol books (catalogued as G2 and G3 at the IGM) are still extant and document the effects of substances on himself (as "ego"), on his co-workers (by name), or as recorded in other sources (such as Cullen). On this topic, see Lucae ("Hahnemanns Prüfungsprotokolle").

15 On Hahnemann's use of them, see Wischner's two-part "Die Benutzung von Repertorien in Hahnemanns Pariser Praxis."

16 Von Hörsten summaries with regard to the *Krankenjournale 1801–03* that the ten most common symptom areas are: disposition, circumstances affecting the spirit (and hence illness), body type, pain, temperature and sweat, sleeping patterns, external aspects such as a skin outbreak, intolerance to foods, symptoms arising from the misuse of pleasurable substances, and the reaction of the pupils (61).

17 "Even at the Universities they were either not known or not applied, although percussion had been discovered and made public by Auenbrugger in 1761 and auscultation by Laennec in 1816" (Haehl 1: 295). By the end of his time in Koethen, Hahnemann was in the possession of a stethoscope.

18 In the commentary volume to the case journals of 1801–3, von Hörsten lists the names of the diseases mentioned along with their frequency, totally 232 references.

19 In addition to Rothschuh's *Konzepte der Medizin*, on the topic of eighteenth-century diagnostic practices, see Rudoph and Henne.

20 Broman (82–3) refers to standard medical textbooks in which pathology is closely linked to physiology by Hieronymous David Gaub (1705–80), Christian Gottlieb Ludwig (1709–73), and Giovanni Battista Morgagni (1682–1771) that stem from Boerhaave's path-breaking attention to pathology.

21 On Boerhaave's medical system, see Cunningham, "Medicine to Calm the Mind."

22 For more on the debate between the empiricists and the rationalists, see Coulter and Rothschuh (*Konzepte der Medizin*). Hufeland, incidentally, also belonged to the empiricists, as Josef Neumann notes in terms that would equally apply to Hahnemann: "Nevertheless, he is aware of the fact that he knows neither the cause of illness nor the mechanisms of causal effect of the applied remedy. Hufeland attempts to compensate for this lack of causal explanation by systematically collecting and comparing therapeutic experiences to assess the effect of therapeutic measures on an as broad as possible basis of experience" (1: 346).

23 See Henne 284. It has also been noted that Hahnemann's teacher in Vienna, Anton Störck, also developed a notion of similitude by matching symptoms of sick patients with those tested on a healthy person. He also tested small amounts of poison hemlock on himself, though not in the homeopathic sense of provings but to ascertain if it was safe for patients (see Mure; Bayr 24–5).

24 Hess points out in two superb essays that Hahnemann is not only indebted to the semiotic focus of eighteenth-century medicine: because

Hahnemann also criticized the classification of the nosologists and came up with his own mode that reduced disease to the mere complex of symptoms (by which to calculate the medical cure), he thereby narrowed the operative semiotic field. For more on Hahnemann's relationship to eighteenth-century medical semiotics, see Henne. On eighteenth-century medical semiotics based on external observation and comparison as opposed to nineteenth-century diagnostics, see Eckart; and Rudolph.

25 Hahnemann's concept of disease shifts, however, starting in the 4th-edition *Organon* of 1829 where he defines disease as "the purely dynamic aberrations of the vital powers" (136, §70).

26 Hahnemann directly opposed humoral practice: "The doctrine of bad humours long enchained mankind" ("Three Current Methods of Treatment," *Lesser Writings* 537; see also "On the Value of the Speculative Systems of Medicine," ibid. 493).

27 Broman refers to Hufeland's *System der praktischen Heilkunde* (2nd ed., 1818) to indicate the two-part process of medical diagnosis standard for the day – naming the disease and specifying its etiology (114).

28 See also Bergengruen's subchapter "Natürliche Signaturen" in *Nachfolge Christi*, where he discusses Paracelsus alongside Ficino, Croll, and other Renaissance thinkers on the signature of things. For instance, he writes that Giovanni Pico della Mirandola entertained the idea, independent from astrology, that philosophers were capable of seeing God's invisible secrets through the visible signs of nature (167). On signs in Paracelsus, see also Böhme, "Denn nichts ist ohne Zeichen."

29 Gantenbein focuses on several parallels rather than fundamental differences between Hahnemann and Paracelsus in order to suggest that Paracelsus functions as a dark Jungian shadow to the founder of homeopathy, who resisted acknowledging direct influences. Haehl, by contrast, says there is no suspicion of Hahnemann having the same ideas as Paracelsus (1: 273). Indeed, Hahnemann speaks out clearly against alchemy ("Aesculapius in the Balance," *Lesser Writings* 421 and "Three Current Methods of Treatment," ibid. 546). Hufeland considered Paracelsus to be a charlatan (see Pfeiffer 99).

30 "Unsinnlich," meaning not dependent on sense or perception, also evokes the term *Unsinn* or nonsense and absurdity.

31 Just devotes a short section to this essay by Benjamin in *Der Akt der Ähnlichkeit*.

32 "A BwO is made in such a way that it can be occupied, populated only by intensities. Only intensities pass and circulate . . . [It is defined by]

dynamic tendencies involving energy transformation and kinematic movements" (*A Thousand Plateaus* 153).

33 The unique symptom must not be confused with the "ungenuine, accidental symptom" that can be caused by a medicine prescribed for a patient that will cloud an accurate semiotic profile ("The Medicine of Experience," *Lesser Writings* 446).

34 I am adapting the terms that Szondi used: he coined the term normative poetics, setting it in opposition to speculative poetics.

35 On *Witz*, see also Lacoue-Labarthe and Nancy (52–8 and 164); Menninghaus (184–6); Gerhard Neumann (452–68); and Schnyder (148–9).

36 For a recent critique of the Romantics' endeavour to espouse irony and dissimilarity in analogy, see Stafford.

37 Wetzels (123) notes that "Ritter once described the analogue process as 'the method of equation' . . . That means that in all phenomena one is looking for similarities, identities, in the hope that one can ascertain a common structure underlying all individuality."

38 For more on how various philosophers rejected logical ratiocination on the basis of analogy, see Stadler (89).

39 On analogy – and its related concepts of affinity and combinatorics – as poetic principles in Romanticism, see Bergengruen ("Magischer Organismus" and "Signatur"); Böhme; Chaouli; Fromm; Gaier; Nakai; Neubauer (*Symbolismus*); Rommel (*Novalis*); Stadler; and Stafford. On analogical concepts in Goethe see the volume edited by Schrader and Weder. On the "Analogieschluss" in natural philosophical medicine, see Rothschuh, "Naturphilosophische Konzepte."

40 Herder wrote: "What we know, we only know from analogy, from the creature to us, and from us to the Creator" (665). Jutta Heinz (35) remarks how close Herder is to the Romantics in terms of the role of analogy and subjective experience instead of objective observation. See also Irmscher (207–35) on Herder's analogical thought.

41 Novalis mentions the "signature of things" (2: 500, #143) and "doctrine of alternating representation of the universe" (2: 499, #137). Not only man but the entire universe spoke (2: 500, #143). A comparison invites itself here to Paracelsus: "Nothing exists for which nature has not provided a sign, and through such signs we can discern the essence of the signified . . . Human beings on earth learn everything . . . through exterior signs and parable, the same is true for every property in herbs, and everything concealed in stones" (366, 368). Also to Jacob Böhme: "All of the exterior visible world and its characteristics are expression or symbol of the interior spiritual world; everything in the interior, including its effects, corresponds to its exterior character" (6: 96).

Chapter Two

42 Schelling similarly wrote that "every single thing represents the universe after its own fashion" (*Bruno* in *Werke* 3: 163/IV: 267).

1. Nicholas Jewson uses the term "beside medicine," as contrasted with the later "hospital medicine" and "laboratory medicine." Michael Stolberg (*Experiencing Illness*) confirms, "Much more than today, patients could also expect a physician to tailor his treatment to their individual bodily constitution and lifestyle" (64).
2. On autobiographical writing at the time see Kuhn; Nussbaum; Smith and Watson; and Stelzig.
3. See also Jane Brown on Goethe's experiments on how to think through the concept of subjectivity.
4. Hahnemann writes in *Aeskulap auf der Wagschale* in 1805: "Galen devised a system for this purpose [to discover the hidden causes of diseases], his four qualities with their different degrees; and until the last hundred and fifty years his system was worshipped over our whole hemisphere, as the *non plus ultra* of medical truth. But these phantoms did not advance the practical art of healing by a hair's breadth; it rather retrograded" ("Aesculapius in the Balance," *Lesser Writings* 421).
5. On regulating the flows in the body and preventing obstruction and stagnation, whether a question of sweat, menstrual blood, excrement, etc., see Stolberg, *Experiencing Illness* 79–156.
6. Thoms investigates such clinical trials conducted on homeopathy between 1820 and 1840. She points out that Hahnemann would have objected to such trials in the first place because they do not focus on the individual.
7. Ironically, although Hahnemann was not a physician who developed an expertise in treating certain organs, he did brand himself as a specialist by creating a new type of treatment. One can say that by the 1830s and 1840s he was very much a reflection of medical specialization.
8. See Stolberg's subchapter on "The Doctor-Patient Relationship" (64–76). On this shift from individualized patient care to focus on the disease, see also Hess ("Diagnose").
9. On clinical trials in dispensaries and military hospitals, see Maehle 268 and 289.
10. On the history of specifics, see Maehle 28 and 286–7.
11. As Wiesing points out, in the eighteenth century for every disease there existed countless remedies and, in turn, every remedy was indicated for numerous diseases (47).

12 Still, Maehle (28) points out that by this time compound preparations were falling out of favour, for the new pharmacopeia preferred the few tried and tested simples.
13 Schelling wrote a scathing, short review of this last essay in Andreas Röschlaub's journal *Magazin zur Vervollkommnung der Medizin* 6 (1801): 221–4. It is reprinted in Schelling, *Historisch-Kritische Ausgabe*, in *Werke* 10: 281–4.
14 On the influence of John Brown on Schelling, see Engelhardt ("Schellings philosophische Grundlegung der Medizin"); Gerabek; Jantzen; Krell; Leibbrand; Rajan; Risse (*History of John Brown's Medical System*); Rothschuh ("Naturphilosophische Konzepte der Medizin"); Tsouyopoulos ("The Influence of Brown's Ideas in Germany"); Wallen; and Wiesing. Recently, Lohff (140–9) and Gerabek (309–33) reverse the notion that Schelling was purely speculative in his writings on nature and medicine. They stress the pre-eminence Schelling lends to empirical verification. For instance, in 1799 Schelling wrote: *"There is absolutely nothing that we initially know except through and via experience [Erfahrung]"* (*Werke* 2: 278).
15 Hahnemann refers derogatorily to medical "systems" in §§1, 54, 55, and 60 of the *Organon*.
16 Tsouyopoulos identifies, however, a second wave of interest in Brown arising around 1813 and culminating during the years 1815–30 due to the popularity of the French physician François-Joseph-Victor Broussais (1772–1838), whose work on inflammatory processes was also based on Brown ("The Influence of Brown's Ideas in Germany" 66).
17 Schmidt notes principal similarities between Brown and Hahnemann: they both rejected earlier nosological classification of diseases, their treatment in bloodletting, and the focus on localizable diseases in favour of a vital force determining the organism as a whole ("Entstehung, Verbreitung und Entwicklung" 47). One can add that both systems attempted to rebalance the entire body via excitability. For a more in-depth comparison of Brown and Hahnemann, see Schwanitz, *Homöopathie und Brownianismus*. For an 1826 article by a follower of homeopathy contrasting Brown as a fanciful systematizer to Hahnemann as an empirically grounded practitioner, see Rummel.
18 For specific studies on Novalis's fragments on medicine, see Engelhardt ("Novalis im medizinhistorischen Kontext"); Fischer; Krell; Neubauer (*Bifocal Vision*); Schipperges; Sohni; and Uerlings (166–78).
19 Hahnemann similarly wrote: "The actual number of genera and species of sporadic and epidemic fevers is probably much greater than is laid down in the works on pathology and nosology" ("Some Kinds of

Continued and Remittent Fevers," *Lesser Writings* 329). Georg Bayr, in his investigation of Hahnemann's self-testing with Peruvian bark, notes that fever was also conceived as a higher pulse rate, as well as alternating hot and cold conditions. The effects of coffee, arsenic, and even pepper could be seen as fever-like inducing (20). On fevers, see also Stolberg, *Experiencing Illness* (144–9); and Broman (114–15). The clinical thermometer registering a body temperature within five minutes was not invented until 1866.

20 On relative dosaging as a novel concept in medicine around 1800, specifically in John Brown, see Wiesemann (146–7).
21 The concept that illness and health were subjective and relative was not uncommon. Carl Arnold Wilmans was another physician who claimed as much in 1799 (see Wiesing 96).
22 In 1810 Friedrich Christian Bach likewise maintained that infectious diseases manifest themselves differently from one individual to the next (5). The early Reil maintained that health was relative and each individual was healthy in her or her own way ("Von der Lebenskraft" 91). Hahnemann is thus not alone in proposing the uniqueness of disease in every individual.
23 On the beginnings of pharmacology, see Maehle.
24 On the rise of the genre "case history" in the second half of the eighteenth century, see Dickson et al.; and Düwell et al.
25 Other chronic illnesses could arise from long-term use of allopathic medicine (*Organon* 138–139; §74), from the privation of such necessities as a healthy dwelling, exercise, meaningful preoccupation, and nourishment, or from protracted alcoholic abuse (*Organon* 139–140; §77).
26 Wischner traces this development in his book *Fortschritt oder Sackgasse?*
27 Dinges notes how unusual consultation of family members was at the time, given that physicians preferred to avoid unwanted lay suggestions for treatment ("Hahnemanns Falldokumentation" 1357).
28 On the individualization of the medical body at this time, Duden writes that the new body had a central position in the self-understanding of the bourgeois class and was a "natural symbol" in which the individual was embodied (28).
29 Broman summarizes the picture: "Thus by 1800 'medical theory' had become 'theories,' a welter of systems and proposals for systems and interpretations of nature published by physician-writers seeking to make a name for themselves" (101).
30 For more on patient letters at the time see the two articles by Stolberg, "'Mein äskulapisches Orakel!'" and "Patientenbriefe in vormoderner

Medikalkultur." On the history of medical case taking, see Gafner et al.; and Geyer-Kordesch, "Medizinische Fallbeschreibungen."
31 This even at a time when, as Stolberg points out, "much more than today, patients could ... expect a physician to tailor his treatment to their individual bodily constitution and lifestyle" (*Experiencing Illness* 66). See Dinges ("Hahnemanns Falldokumentation" 1359–60) for a comparison between the number of patients Hahnemann and other physicians saw in a day and to what degree they noted the patient's narrative. For several graphs on the number of patients, including new ones, Hahnemann would see monthly during his years in Leipzig, as well as a breakdown of patients according to gender, age, occupation, and residency, see Schreiber.
32 On self-surveillance via dietetic and hygienic regimens around 1800, and hence normalization of the body, see Brockmeyer; Egger; Dreißigacker; Koschorke, *Körperströme*; Mahler; Sarasin; Thums; Vigarello; and Zumbusch. Sarasin writes that there are two medical models that establish themselves in the Enlightenment – those of irritability and of the vital life force. The *souci de soi* is dependent on the successful self-regulation of the nerves and emotions, whereas the model of health as balance of the vital life force focuses more on conscience and norm regulations, with direct effects on the self-perception of one's body (211–12). Apart from Brockmeyer, homeopathy is hardly mentioned in these studies.
33 With its clear system laid out in the *Organon*, which patients could read on their own, homeopathy eventually allowed and even encouraged lay self-medication. See Baschin, *Die Geschichte der Selbstmedikation*.
34 On the topic of compliance, see Dinges, "Hahnemanns Falldokumentation" 1357. Brockmeyer wisely points out that dietetics is a site not only of self-regulation but also of both subordination and resistance (127).
35 Busche lists the works of Hahnemann that deal with dietetics: "Diätische Gespräch mit meinem Bruder" (1792), "Abhärtung des Körpers. Erstes Fragment" (1792), "Sind die Hindernisse der Gewissheit und Einfachheit der practischen Arzneykunde unübersteiglich?" (1797), "Der Kaffee in seinen Wirkungen" (1803), and "Aeskulap auf der Wagschale" (1805).
36 A 1991 study by Thomas Genneper also investigates what it was like to be Hahnemann's patient.
37 Although one cannot yet speak of a Cousinian immaterial self "given to its possessor whole and a priori" (Goldstein, *The Post-Revolutionary Self* 6), Hahnemann seems to anticipate, in addition to Freud's talking cure, Victor Cousin's (1792–1867) program of self-talk. What one does not find in Hahnemann, however, is a Freudian suspicion of language, i.e., that

language conceals more than it reveals. Instead, Hahnemann's notation of the patient's exact wording and his subsequent matching of these symptoms to those recorded in his *materia medica* exhibit his eighteenth-century belief in the semiotic transparency of language. On a more theoretical level, as Séverine Pilloud observes in reference to patient letters, there can be no principal distinction made between experience and its narration, because the former can only be given form through words (46).

38 On their correspondence, see Inge Christine Heinz. See Schriewer on another female patient, who, Schriewer observes, uses her correspondence with Hahnemann as a form of intellectual working through of her illness not otherwise permitted in her circle of family and friends.

39 Monika Papsch, however, says that in contrast to 1821, by 1833–5 one cannot say that Hahnemann was systematically testing on his patients. Instead he demonstrated hesitation about which remedy or dosage to choose (134–5).

40 Von Hörsten similarly points out that between 1801 and 1803, half of the patients did not return after the first consultation, for which the case journals offer no explanatory reasons (52).

41 Baschin (*Ärztliche Praxis* 228) refers to similar blind spots in Bönninghausen's praxis: (1) his motivations and reflections are unknown, (2) there is no explanation for the dwindling number of patients, nor (3) for why some came only once. Most importantly, (4) there is no documentation of the decision-making process.

42 On Bönninghausen's published case history of Annette von Droste-Hülshoff, see the article by Dinges and Holzafpel.

43 On a related note, for the development in the history of homeopathy of a remedy specific to one's constitutional type, see Czech.

44 See also Geyer-Kordesch, "Georg Ernst Stahl's Radical Pietist Medicine."

45 Carsten Zelle lists among the discursive elements that formed this new eighteenth-century discipline: methodological reliance on empirical experience, experimentation and observation, focus on the individual case and case studies, and the education of the public in self-observation and the ethics of self-care (209). All these elements pertain to the birth of homeopathy. On the scientific and social history of feeling, see the excellent volume by Frevert, in particular the essay by Bettina Hitzer on the medical context.

46 For recent work on Lavater's science of physiognomy, see Gray; Lyon.

47 For a lengthier discussion of Bach in terms of the fear of contagion, see Zumbusch 60–6.

48 In a recent article, Fritz Breithaupt discusses the development of psychology, involving the topics of memory, recollection, and trauma, in terms of the rise of Romantic selfhood. His approach does not consider, however, psyche–soma relations and hence brackets out discussion of the all-important state of medicine around 1800.
49 On the dominant discourse of nerves, see, in particular, part 3 of Stolberg, *Experiencing Illness*.
50 Vila writes: "Sensibility was, moreover, generally cited as the cause underlying the pervasive and troublesome condition of vapors in worldly women and men, who, out of their extreme susceptibility to the slightest irritant, suffered from hypochondria, hysterical paroxysms, or at the very least, poor digestion and enfeebled offspring" (46). On hypochondria in eighteenth-century German life and letters, see Potter.
51 See also Stalfort and chapter 2.3.3, on "Von der Macht des Gemüths," in Egger. Stalfort, however, sees the term *Gemüt* dwindling in the nineteenth century, replaced more and more by feelings (*Gefühle*). On the longer trajectory of the shift in vocabulary to describe psychological states, especially in the British tradition, see Dixon.
52 I do not wish to dispute, however, the presence in the long eighteenth century of a debate about the influence of the mind on the body. See, for instance, Wright; and Rey, "Psyche, Soma." Brockmeyer therefore places Hahnemann squarely within the eighteenth-century legacy of the unity of body and mind, which she links as well to humoral pathology (215).
53 Summaries of Hahnemann's notations regarding *Geist* and *Gemüt* can be found in the commentary volumes to the *Krankenjournale D34 (1830)*: 64–5 and *D38 (1833–35)*: 64–7.
54 The phrase Hahnemann uses here is "Gemüts- und Geisteszustand." Throughout the *Organon* he frequently deploys *Gemüt* and *Geist* (mind/spirit) together as a pair and did so from the first edition onward.
55 Reil is becoming the focus of more scholarly investigation. See Koschorke, "Poiesis des Leibes"; Speler; Steger; Rieger.
56 Because Reil believed corporeal causes could underlie psychic disorder, Rothschuh classifies him as a "Somatiker," different from someone like Heinroth, the "Psychiker" (*Konzepte der Medizin* 312). On this distinction see also Kutzer, "'Psychiker' als 'Somatiker.'"
57 See Borgards; Kaufmann; Kutzer ("Stimulation"); Luyendijk-Elshout; and Porter ("Barely Touching") on the repressive psychic cures of the period.
58 For more on Pinel's moral treatment see Goldstein, *Console and Classify*.

59 For a non-therapeutic and hence contrasting view of the history of psychology between 1700 and 1840, see Matthew Bell. Bell interprets its development in terms of philosophy and literature.
60 Borgards makes these statements in reference to Marc-Antoine Petit's *Discours sur la douleur* (1799) and Carl Anton Bitzius's *Versuch einer Theorie des Schmerzens* (1803). For further studies on the history of pain see Morris; Rey, *History of Pain*; Stalfort; and Tanner.
61 For more on the history of self-testing in homeopathy, see Bayr; Schott; Walach ("Methoden"); and Wettemann's edition of Hahnemann's *Fragmenta*.
62 See Bayr; Gantenbein ("Der Einfluß"); and Mure.
63 In particular, see Brockmeyer (esp. 215); Hess ("Hahnemann und die Semiotik"); Große-Onnebrink; and Schmidt (*Die philosophischen Vorstellungen*). In the contemporaneous debates on whether the physician should be a practitioner of the art of healing or an academic who investigated systems of pathology and nosology, Hahnemann certainly aligned himself, along with Hufeland, as the former. In that sense, he would have called himself an empiricist. On this debate, see Wiesing.
64 Hahnemann's own use of the term *Empirie* cuts two ways: on the one hand, he referred to the "grossest empiricism" of single remedies being prescribed for every symptom ("Cure and Prevention of Scarlet-fever," *Lesser Writings* 374) and to the "empiricism and superstition" (*Gesammelte kleine Schriften* 428) of the use of folk remedies. Here empiricism refers to uninformed, atomistic observation. On the other hand, he also refers to the "para-empiricism" of Brown, which stands for "the evil demon" not based on bedside know-how but on sheer speculation, while empiricism means "the good genius of experience" ("Three Current Methods of Treatment," *Lesser Writings* 522). Section 67 of the *Organon* refers to the "incontrovertible and self-evident truths which nature and experience (*Erfahrung*) have laid before us" (132). The second edition of Adelung's dictionary of 1793 associates experience (*Erfahrung*) with "knowledge attained through the senses" (1: 1888). Grimm's dictionary cites *Erfahrung* in conjunction with Kant's usage as synonymous with empiricism (3: 793–4).
65 That Novalis could write "idealism is nothing but genuine empiricism" (2: 550, #402) alone indicates how "empirical" at the time resonates with Kant's doctrine of transcendental idealism, i.e., that we know things, not in themselves, but through how they appear to us. In his chapter "Experience and Epistemology: The Contest between Empiricism and Idealism," Martin Jay locates the philosophical struggle already in Locke: "If both

everything *in* the mind is experience and everything in the mind *arises* from experience, then experience is just another word for the contents of the mind and fails to explain very much of anything" (56).
66 See also Van den Berg.
67 Solhdju mentions Hahnemann in a note as conducting self-experimentation, but solely his initial study of Peruvian bark (9).
68 To the same effect, in writing about Romanticism and the natural sciences, Gabriele Rommel concludes: "It was only in the second half of the nineteenth century that experimental research and speculative natural philosophy became strictly differentiated, for until this point they had been closely intertwined in their attempts to determine the relationship between nature and spirit" ("Romanticism and Natural Science" 213). Classical homeopathy cannot anachronistically be seen to predate this division and portrayed as solely committed to objective research.
69 Where Hahnemann differs, however, from the eighteenth-century botanical atlases with their idealized, perfected illustrations is that, unlike them, he does not extract the typical from the wealth of natural particulars, but hones in on the unique. He thus combines the eighteenth-century dedication to the accuracy of a draftsman with nineteenth-century curiosity about individual lives, their variations, and anomalies.
70 There has been next to no scholarship comparing Goethe and Hahnemann, however. An exception is a 1947 article by Rudolph Tischner (1879–1961) ("Hahnemann und Goethe") in which the author (the parapsychologist who coined the term "extrasensory perception" and who wrote extensively on homeopathy) concludes that Hahnemann, as an empirical scientist and Enlightenment thinker, shares little with the poet. An understandably different approach is taken more recently by the Germanist Robert Tobin, who conceptualizes a homeopathic principle for analysing the calculated educational, corrective moves of the Society of the Tower in Goethe's novel *Wilhelm Meisters Lehrjahre* (*Wilhelm Meister's Years of Apprenticeship*, 1795–6). In a recent dissertation on the figure of the physician in Goethe's *Faust*, Buchinger mentions Hahnemann only briefly.
71 On Goethean scientific experiment, see in particular Hartmut Böhme ("Lebendige Natur"); Egger (chapter 1.3.1 on "Goethe und das Experiment"); Erpenbeck; the essays in the collections edited by Amrine et al.; and Seamon and Zajonc. See John on Goethe's concept of materialism.
72 Daiber concludes that for Novalis every experiment on external nature is also an inner experiment (*Experimentalphysik des Geistes* 110). The emphasis

is hence on self-exploration. See also Nassar, "'Idealism is nothing but genuine empiricism.'"
73 For more on Ritter's empiricism, see Klaus Richter.
74 In 1820 Johannes Evangelista Purkinje ventured into dangerous self-experimentation with vertigo. On the history of vertigo see Janz et al.
75 For recent studies on Ritter's poetic science, see Bergengruen ("Magischer Organismus"); Henderson; Holland; Lothar Müller; Joan Steigerwald ("Figuring Nature"); Strickland; and Wetzels.
76 Levin cautions the historian against the temptation to diagnose retrospectively in "Krankheiten – historische Deutung versus retrospective Diagnose."
77 The most salient example of this approach in the history of homeopathy is by Brockmeyer.
78 Friedrich Schlegel also wrote that the classical work must never be fully understandable (*Kritische Ausgabe* 2: 149, #20). In an appeal to the reader, he also stated that even the most universal, complete works of poetry and philosophy appear to avoid final synthesis (2: 255, #451). For further discussion of the interminable act of reading in German Romanticism, see my book *Delayed Endings* (122–32).
79 For more on the tradition of the physician as genius as it arose in the eighteenth century, and as cultivated in particular by Johann Georg Zimmermann, see Dinges, "Medizinische Aufklärung." Dinges writes that this physician of the Enlightenment serves as a good example of the ambivalence of Western modernity in which knowledge is inseparable from power and that the exercise of power always hangs together with the state of knowledge (150). In *Kunst oder Wissenschaft* Wiesing focuses on the issue of whether the physician was to be regarded as an artist or scientist, although "artist" in this context means the insightful, gifted practitioner.
80 In this respect, Hahnemann resembles Humphry Davy, who paraded attributes of a daring scientific genius who risked self-experimentation (see Golinski).

Chapter Three

1 On the history of the Q-potency, see Jütte (*The LM Potencies in Homeopathy*); Mayr; and Sauerbeck.
2 Tischner points out that Hahnemann first uses the word "dynamic" in 1800 (see "A Preface," *Lesser Writings* 346). He also uses the word "virtual" in opposition to "chemical" ("Goethe und das biologische Grundgesetz" 361).

3 Translated by A.S. Kline. http://www.poetryintranslation.com/PITBR/German/FaustIIActIScenesItoVII.htm.
4 Hahnemann was interested in the sickness mercury caused, "in order to show how opium can cure it, by virtue of similarity of action" ("The Curative Powers of Drugs," *Lesser Writings* 286).
5 See my discussion in chapter 1 of opium. Hufeland, incidentally, was against opium's overuse, although he did prescribe it (see Pfeiffer 160 and 164). In particular, he was cautious about prescribing drugs for children (Pfeiffer 123).
6 For more on Novalis's dynamic *Lebenskunstlehre*, intensifying rather than regulating body and spirit, see Uerlings (172–8); Sohni; and Thums.
7 On Burdach, see Sarasin 214–17.
8 Novalis brilliantly remarks that even dilutions have the potential to stimulate, much like misery or deprivation (2: 608, #594).
9 Another example of Hahnemann subscribing to the notion that "poison" is a relative concept is his division into primary and secondary effects (Tischner, *Geschichte der Homöopathie* 685). For more on the concept of "relative" in dispensing medications, on the subjectivity of well-being in Brown and Röschlaub, and on the differing susceptibility between individuals to infection and illness, see Wiesemann. Her main argument is that only with the notion of relative health can the notion of addiction be developed, because there is a separation between *how* you feel (good on opium) and from *what* you are (addicted).
10 For an excellent conceptual medical history of *Reizbarkeit* and *Reiz*, see Möller. Whereas Haller used the term *Reizbarkeit* in 1752 to refer to the reaction of the muscles, by 1796 Reil was using it as a synonym for the vitality of the entire organism. Hufeland uses the term *Erregbarkeit* in 1800 as the capacity of the whole organism to register and respond to *Reiz*. Hahnemann can be said to follow suit in speaking about the synergistic responsiveness of the entire body to the homeopathic remedy. The distinction from Brown is crucial here: whereas Brown maintains that an organ must be stimulated to responsiveness through an external source (with the body being passive), beginning with Reil, *Reizbarkeit* signals the body's vital receptivity to stimuli (Möller 32).
11 In the considerable literature on Novalis and Romantic medicine, the references to Hahnemann are surprisingly few and far between. Uerlings briefly notes a similarity (176), as does Engelhardt ("Novalis im medizinhistorischen Kontext" 77). As Novalis's medical notations were not published for the most part until the beginning of the twentieth century (Uerlings 167), it is not a question of direct influence.

12 Unlike Wischner, I shall not aspire to determine which elements in Hahnemann's thought led to dead ends and which were medically productive.
13 For more precise information on the preparation of a drug (research, source, and manufacture, including dilution, trituration, and succussion) and its dispensing (when and how much to take, choice and modification of potency, length of effect, reaction to, repetition of dose, course of treatment), see Haehl 1, chap. 24; Mayr; Papsch's commentary on D38; and Wischner, *Fortschritt oder Sackgasse?* (167–270). Jütte (*Samuel Hahnemann*) offers a detailed summary in English.
14 He appends a note stipulating that in "the treatment of chronic diseases, even after the complete restoration of health, [it is necessary] to continue giving for some months longer a small quantity of the same medicine that cured the disease, but at ever longer and longer intervals, in order to eradicate every trace of the chronic disease in the organism that has been for years accustomed to its presence" ("The Medicine of Experience," *Lesser Writings* 455).
15 On Haller, his relationship to the Montpellier school, and other eighteenth-century responses to his theory, see Steinke; and Vila.
16 As Botsch phrased it, "questions also remained open about the theory of a physical *Lebenskraft*: did *Lebenskraft* reside in certain elementary elements, was it dependent on the form of matter, the organization of its atoms, or was it an additional principle to enhance matter?" (236).
17 Steigerwald, "Treviranus' Biology" (107). See also her other essays on the topic of vitalism, "Rethinking Organic Vitality" and "Instruments of Judgment."
18 English scholarship on late-eighteenth- and early-nineteenth-century vitalism concentrates on theories of embryogenesis, morphological processes, and *Bildungstrieb* (growth impulse or formative drive), leading up to Darwin's evolutionary thought. For instance, Kelley, Miller, and Richards are among the many who have investigated the inner-directedness and inner purposiveness of plants in Goethe. Even though Hahnemann does not talk about generation or morphological development, he does resemble Goethe and Treviranus in that, like them, he posits a kind of germ or kernel that is transferred unseen as energy. This kernel of energy is what resides, enhanced, in the infinitesimal toxinogenic remnant. For a recent collection that looks broadly at the influence of the *Lebenskraft* debate in German science and arts, see McCarthy et al.
19 The scholarship in German on medical *Lebenskraft* is more extensive: see Arz; Botsch; Engels; Goldmann ("Von der Lebenskraft"); Lohff ("Zur Geschichte der Lehre von der Lebenskraft"); Neuburger; Noll; and

Rothschuh (*Physiologie*, chap. 11). On the French medical tradition of vitalism, see Rey, "Psyche, Soma"; and Williams.

20 Another current fascination with vitalism is reflected in the popularity of works by Jane Bennett and Timothy Morton. By "vibrant matter," Bennett means a kind of vibrationality within nature or an energy of things, together with the invisible flows and sympathetic correspondences between them. Morton calls it "the infinite being of things" (22). The discursive constellations around 1800 into which homeopathy can be placed promise to contextualize this scholarship historically.

21 Without going into detail, Johanna Geyer-Kordesch closes her essay on Stahl's Pietist medicine by saying that "his championing of the imagination as the creative element in the psyche's holistic construction of perception is most certainly important for Romanticism" (87).

22 On the influence of Haller's sensibility on eighteenth-century French medicine, ethics, and literature, see Vila.

23 Coulter, though, sees Hahnemann as culminating a lineage (Paracelsus, Stahl, van Helmont) of what he calls "empirical physicians" who operated according to a "sense-based epistemology and vitalistic physiological assumptions" (xiii). Coulter formulates this "schism in medical thought" between the (good) empiricists and the (bad) rationalists who "sought medical certainty in the rules of formal logic" (viii) in order to place homeopathy at the pinnacle of dedication to therapeutic practice.

24 Hahnemann greatly admired Hufeland, writing in 1801 that his "soul is animated by truth alone!" ("View of Professional Liberality at the Commencement of the Nineteenth Century," *Lesser Writings* 363). For studies on Hufeland, see Egger; Goldmann (*Christoph Wilhelm Hufeland im Goethekreis*); Josef Neumann; Pfeiffer; Rothschuh (*Physiologie* 330–6); Schwanitz (*Die Theorie*); and Zumbusch.

25 At the same time, Broman is correct to point out that "one lingering legacy of Brunonianism after its disappearance as an organized movement was the tendency of German physicians to emphasize illness as a dynamic problem affecting the entire organism, rather than a localized and material affliction" (156). And Mocek has argued that Brown's teachings led to the rethinking of the concept of organic matter – as the need for a science of life (91).

26 In general, on the topic of the Brunonians' belief in the natural healing powers, see Wiesing: "Nature becomes reasonable, curative, she transforms into genius . . . into the actual agent, therapeutically ingenious and omnipotent" (297). On the role of natural healing in the eighteenth century and its earlier tradition, see Duden (194–201). On its role in Hufeland, see Josef Neumann (350–4). On the later nineteenth-century development of the *vis medicatrix naturae*, see auf der Horst.

Finally, for a more extensive historical overview regarding the healing powers of nature, see Neuburger.
27 To be sure, Hahnemann also attacked those physicians who advocated the use of purgatives and emetics, which were seen to imitate the body's own natural evacuative processes (see "Introduction," *Organon* 40–50 and "The Medicine of Experience," *Lesser Writings* 439).
28 As Schmidt points out, Hahnemann did not, though, explicitly situate himself within the long tradition of *Lebenskraft* ("Die Entstehung" 49). On the role of *Lebenskraft* in Hahnemann, see Haehl 1, chap. 22.
29 In his assault on homeopathy, Johann Stieglitz (1767–1840) noted in 1835 how unconventional Hahnemann's linguistic usage was: the term "dynamic-spiritual" is applicable to activities of the soul, not to *Lebenskraft* (123).
30 Compare this to Reil, who much earlier in 1795 spoke of the false tuning ("Mißstimmung" and "falsche Stimmung") of *Lebenskraft* as being the most common cause of disease. The tuning of *Lebenskraft* is variable, dependent on the quality of delicate matter ("feine Stoffe"), and can be restored, for instance, by removing the cause of irritation ("Von der Lebenskraft" 93). Reil and Hahnemann thus differ not in seeing the need for retuning *Lebenskraft*, but in how this process is to be undertaken.
31 Wischner also sees Hahnemann aligning *Lebenskraft* in the human organism with the vitalistic equivalent in all of nature (*Fortschritt oder Sackgasse?* 72).
32 One should add that by 1800, "[the] notion that the vital principle was not only associated with the organic but also inorganic matter oriented itself in terms of the *Naturphilosophie* of Friedrich William Joseph Schelling" (Botsch 171). Botsch also quotes the physician Gottfried Christian Reich (1769–1848), writing in 1810 that "nothing perishes in the entire cosmos, nothing is added to it, and everything which presents itself as phenomenon is only alteration in the form of living matter" (170).
33 Although the *Reine Arnzeimittellehre* mentions primarily botanical substances, later in life Hahnemann devoted himself more to the investigation of minerals, especially for chronic diseases (Wischner, *Fortschritt oder Sackgasse?* 168–9). He also for the first time uses an animal product, sepia.
34 One of Wischner's main findings is that in the third *Organon* and increasingly after the fifth Hahnemann explains *Lebenskraft* itself as the cause of symptoms (146). This is because the illness marks the healing attempts of *Lebenskraft* (335).
35 Apropos of the medical uses of electricity, Hahnemann writes in 1807 that experiments with electricity indicate how one can overcome pain

by producing a stronger current. He offers this as proof of homeopathic healing (*Gesammelte kleine Schriften* 464).

36 For other references in Hahnemann to Kant, see "On the Value of the Speculative Systems of Medicine" *Lesser Writings* 496 and "Spirit of the Homoeopathic Doctrine of Medicine" *Lesser Writings* 617. Risse has demonstrated how "Kant's critical writings were beginning to achieve a wide circulation after 1790. He was viewed by physicians as an enemy of dogmatism who was leading human reason back to ... a genuine knowledge of the world of experience" ("Kant, Schelling" 147). But Risse also correctly points out that "in Kant's epistemology the entire world of experience was actually a product of the human mind, which ... ordered the sensations according to its own structure" (147). Therefore, "only systematic unity could elevate ordinary knowledge – for Kant a mere aggregate or 'rhapsody' of notions – to the rank of 'science.' This effort of 'scientific' systematization could only be justified if a basic identity and lawfulness existed in nature behind the apparent complexities" (148). Although Risse does not discuss Hahnemann, his articles go far in explaining why the founder of homeopathy formulated what he saw as "scientific" principles to arrange the empirical data he was amassing.

37 See also her essays "Instruments of Judgment" and "Kant's Concept of Natural Purpose."

38 Just as Hahnemann potentized remedies by diluting them, so too Novalis wrote: "I realize the Golden Age by expanding its polar opposite" (2: 622, #634).

39 One of Eckart Förster's main arguments in *The Twenty-five Years of Philosophy* is that Goethe expands Kant's concept of intuitive understanding. On this development, see also Hindrichs and Amrine.

40 I disagree strongly with Wischner, who sees Hahnemann's increased usage of *Lebenskraft* not to stand in contradiction to his demands for a medicine based on reason. In fact, Wischner jumps to the conclusion that, precisely because Hahnemann's teaching of *Lebenskraft* and other dynamic forces is reason-based, it sets him apart from Romanticism and *Naturphilosophie* (342). Bettina Brockmeyer follows suit and calls Hahnemann's dynamic concept borrowed from vitalism to be "enlightened" and in line with Kant's empiricism (79). Brockmeyer and Wischner are symptomatic of much scholarship on homeopathy that wants to force it into an Enlightenment framework, but without a nuanced understanding of Kantian epistemology.

41 The scholarship on mesmerism is rich. See Bark; Barkhoff (*Magnetische Fiktionen*); Darnton; DeLong; Tatar; Weder; and Winter, to name a few. In particular, see Rieger's conceptualization of magnetism and mesmerism around 1800 in terms of twenty-first-century cybernetic flows and wireless networks. On homeopathy and mesmerism, see Wittern and Eppenich.

42 Hufeland, too, warned against the charlatanry of mesmerism as early as 1784 in an essay entitled "Mesmer und sein Magnetismus."

43 The sexual implication is clear: Hahnemann even said that the magnetizer must "have a very moderate inclination for sexual intercourse" so that the "abundance of the subtle vital energy, which would else be employed in the secretion of semen, is disposed to communicate itself . . . through the medium of the touch, seconded by a strong intention of the mind" (*Organon* 228, §293).

44 "Vaccination" > Latin *vaccinus*, "pertaining to a cow" > *vacca*, "cow." On Jenner's discovery seen in the historical context of natural history, consult Rusnock.

45 Vaccination was compulsory in Bavaria after 1805, in Baden after 1807, in Württemberg after 1818, but not in Saxony, where Hahnemann lived and practised (Heinz and Wischner 181). On religious opposition to vaccination, see Lobo.

46 For a detailed, historic overview of Hahnemann's assessment of cowpox vaccinations, how and when he practised them, and when he recommended an alternative prophylaxis of smallpox by sniffing *rhus toxicodendron*, see Heinz and Wischner. See also Heinz, 192–7.

47 The prophylactic treatment with nosodes was developed by subsequent homeopaths. Nosodes come from a virus taken from humans (sarcodes from animals) and are a form of isopathy, *aequalia aequalibus*. Hahnemann mentions them when he was 79 years old in his *Krankenjournal* of 1833–5 (*D38*). Because nosodes are neither tested on the healthy nor matched to individuals, they fall outside strictly classical homeopathic principles. See Vieracker.

48 Hahnemann does include a footnote in the *Organon* in which he states that testers will not suffer from any detrimental effects to their health: "On the contrary, experience has shown us that they only render the body more apt to repel all natural and artificial morbific causes, and harden it against their influence. The same experience also teaches, that thereby the health becomes more firm, and the body more robust" (*Organon* 168, §141).

49 See Egger; Koschorke, *Körperströme* (64); Sarasin; Tobin; and Zumbusch. Johannes Türk has argued for the continuation into the twentieth century

of this model of pitting and steeling the self against infectious others. See also Kroker et al.; Pias; and Sarasin et al.; and Tauber.
50 See my articles on Novalis. Of special note: Thums records a shift in Novalis and Schlegel away from the dietetic adherence to measure in order to experience alternating extremes.
51 Engelhardt lists "potency" as one of the formal principles organizing Romantic natural science, along with polarity, analogy, metamorphosis, and mathematical principles ("Natural Science in the Age of Romanticism" 120). See also his "Naturforschung im Zeitalter der Romantik" 35–40. Barkhoff also points out that "thinking in polarities, analogies, potentialisations and metamorphoses is characteristic of romantic science and nature philosophy" ("Romantic Science and Psychology" 211). In contrast to Hahnemann, though, in Schelling, Oken, and Henrik Steffens (1773–1845), *Potenzierung* refers to ascending orders in nature. On Novalis and "Potenz," see, in addition to Neubauer, Wasmuth. With regards to polarity, in homeopathy the mediation between polar opposites occurs when like cures like (a toxin heals) and when indiscernible amounts of an ingredient emanate immeasurable energy. In addition, as just discussed, Hahnemann self-tested with magnets.
52 See Mayr on this development. Also Jütte (*The LM Potencies*); Sauerbeck; and Tischner ("Über den Begriff des Dynamischen").
53 These paragraphs are not provided in Constantine Hering's English edition of the *Organon*, hence the translation is my own from Josef Schmidt's critical edition.

Conclusion

1 On Novalis and the language of nature, see in particular Hartmut Böhme ("Denn nichts ist ohne Zeichen") and Goodbody.
2 In 1954 the anthroposophist Theodor Schwenk schematized how plants materialized (*Substanzwerdung*) universal forces or rhythms that could then be exponentiated and released (*Potenzvorgang*) in the homeopathic remedy. The homeopathic medium takes up traces of these rhythmic forces. He explained the simile-enigma as nature imprinting in individual plants specific human organ functions (24).
3 On the relationship between homeopathy and philosophy, see Schmidt, "Homöopathie und Philosophie" and the volume of essays edited by Appell, *Homöopathie und Philosophie & Philosophie der Homöopathie*.

4 The most thorough investigation into science and philosophy in the age of Goethe, as it is subtitled, is the study by Robert J. Richards, *The Romantic Conception of Life*.
5 For essays on Goethe and Spinoza, see Amrine ("Goethean Intuitions"); Lange; and Vlasopoulos.
6 Dalia Nassar entitles her recent book *The Romantic Absolute* and lists the three main questions facing Goethe, Novalis, Friedrich Schlegel, and Schelling as how to relate (1) the mind and nature, (2) the one and the many, and (3) the infinite and the finite (2). See also my discussion of the infinite and finite in homeopathy in terms of calculus in chapter 3.
7 Amrine explains the *natura naturans* as "the 'Gott-Natur' (God-nature) from which the discrete, finished forms of *natura naturata* flow" ("'The Magic Formula We All Seek'" 257). Goethe wrote in 1785 to F.H. Jacobi, "I only recognize [divine being] in and from *rebus singularibus*, whose closer and deeper observation no one more than Spinoza himself encourages, although before his gaze all individual things seem to disappear" (WA 4.7: 63). Schelling, too, saw the activity or productivity of nature in its smallest parts: "Because an infinite product is evolving itself in nature . . ., so must there be thought an infinite diversity of simple actions" (*Werke* 2: 5).
8 According to Deleuze, the plane of immanence allows for "a radical empiricism" (*What Is Philosophy?* 47).
9 Translation taken from Johann Wolfgang Goethe, *Scientific Studies*, ed. and trans. Douglas Miller (Princeton: Princeton UP, 1995), 8.
10 Ibid., 15–16.
11 See Hindrichs on Goethe's *anschauende Urteilskraft*. On Fichte, Amrine, "'The Magic Formula We All Seek.'" As well, Neubauer, "Intellektuelle, intellektuale und asthetische Anschauung."
12 Goethe also speaks of a higher empiricism ("eine höhere Empirie" [HA 12: 437]).
13 "For Spinoza, *scientia intuitiva* is the third and highest mode of knowledge; understanding things in light of intuition was for him 'the highest effort of the mind and its highest virtue.' Its ultimate goal and promise is to reveal 'the knowledge of the union existing between the mind and the whole of nature'" (Amrine, "'The Magic Formula We All Seek'" 252).
14 Schmidt attributes this passage to the influence of Kant's critical philosophy in his annotations to *Organon der Heilkunst* 287.

Bibliography

Primary Works

Adelung, Johann Christoph. *Grammatisch-kritisches Wörterbuch der Hochdeutschen Mundart*. 2nd ed. Leipzig: Breitkopf, 1793.

Baader, Franz von. *Sämtliche Werke*. Ed. Friedrich Hoffmann. Aalen: Scientia, 1963.

Bach, Friedrich Christian. *Grundzuege zu einer Pathologie der ansteckenden Krankheiten*. Halle and Berlin: Buchhandlungen des Hall. Waisenhauses, 1810.

Baudelaire, Charles. *Oeuvres complètes*. Ed. Claude Pichois. Bruges: St Catherine, 1968.

Benjamin, Walter. *Gesammelte Schriften*. Ed. Rolf Tiedemann and Hermann Schweppenhäuser. Frankfurt am Main: Suhrkamp, 1980.

Bichat, Xavier. *Physiologische Untersuchungen über Leben und Tod*. Trans. D. Veizhans. Tübingen: Jakob Friedrich Heerbrandt, 1802.

Böhme, Jacob. *Sämtliche Schriften. Faksimile-Neudruck der Ausgabe von 1730*. Ed. Will-Erich Peuckert. 11 vols. Stuttgart: Frommann, 1955.

Bönninghausen, Clemens von. *Das erste Krankenjournal (1824–1830)*. Ed. Luise Kunkle. Quellen und Studien zur Homöopathiegeschichte 14. Essen: KVC, 2011.

– *Die Homöopathie: Ein Lesebuch für das gebildete, nicht-ärztliche Publikum*. Münster: Coppenrath, 1834.

Carus, Carl Gustav. *System der Physiologie umfassend das Allgemeine der physiologischen Geschichte der Menschheit, die des Menschen und die der einzelnen organischen Systeme im Menschen, für Naturforscher und Aerzte*. 2 vols. Dresden and Leipzig: Gerhard Fleischer, 1838–40.

Coleridge, Samuel Taylor. "Theory of Life." In *Collected Works*. Vol. 11.1, *Shorter Works and Fragments*, ed. H.J. Jackson and J.R. de J. Jackson, 481–557. Princeton: Princeton University Press, 1995.

Diderot, Denis, and Jean le Rond d'Alembert. *Encyclopédie, ou, Dictionnaire raisonné des sciences, des arts et des métiers*. 35 vols. Neufchastel: Faulche, 1765.

Droste-Hülshoff, Annette von. *Historisch-kritische Ausgabe*. Vol 8.1. *Briefe 1805–1838*. Ed. Walter Gödden. Tübingen: Niemeyer, 1987.

Eichendorff, Joseph Freiherr von. *Neue Gesamtausgabe der Werke und Schriften*. Ed. Gerhart Baumann. Stuttgart: Cotta, 1958.

Eschenmayer, Carl August von. "Allgemeine Reflexionen über den thierischen Magnetismus und den organischen Aether." *Archiv für den Thierischen Magnetismus* 1 (1817): 11–34.

Fichte, Johann Gottlieb. *Sämmtliche Werke*. Ed. Immanuel Hermann Fichte. 8 vols. Berlin: Veit & Comp., 1845–6.

Galvani, Luigi. *Commentary on the Effects of Electricity on Muscular Motion*. Ed. Margaret Glover Foley. Norwalk: Burndy Library, 1954.

Goethe, Johann Wolfgang von. *Briefwechsel Goethe und Knebel (1774–1832)*. Vol. 1. Leipzig: F.A. Brockhaus, 1851.

– *Gedenkausgabe der Werke, Briefe, und Gespräche*. Ed. Ernst Beutler. 24 vols. Zurich: Artemis, 1948–54.

– *Werke*. Hamburger Ausgabe. Ed. Erich Trunz. 14 vols. Munich: Beck, 1982–2008.

– *Werke*. Weimarer Ausgabe. Ed. im Auftrage der Grossherzogin Sophie von Sachsen. 144 vols. Weimar: H. Böhlau, 1887–1919.

Grimm, Jacob, and Wilhelm Grimm. *Deutsches Wörterbuch*. Vol. 3. Leipzig: Hirzel, 1862.

Hahnemann, Samuel. *Die chronischen Krankheiten: Ihre eigentümliche Natur und homöopathische Heilung*. Vol. 3. Stuttgart: Haug, 2003.

– *Die chronischen Krankheiten: Theoretische Grundlagen*. Intro. Will Klunker. Vol. 1. Heidelberg: Haug, 2001.

– *"Fragmenta de viribus medicamentorum": Die erste Materia medica homoeopathica*. Ed. Marion Wettemann. Diss., Tübingen, 2000.

– *Gesammelte Arzneimittellehre: Alle Arzneien Hahnemanns: Reine Arzneimittellehre, Die chronische Krankheiten und weitere Veröffentlichungen in einem Werk*. Ed. Christian Lucae and Matthias Wischner. 3 vols. Stuttgart: Haug, 2007.

– *Gesammelte kleine Schriften*. Ed. Josef M. Schmidt and Daniel Kaiser. Heidelberg: Haug, 2001. Eng. abridged version, *The Lesser Writings*. Trans. R.E. Dudgeon. New York: Radde, 1852.

– *Krankenjournale D2–D4 (1801–1803)*. Ed. Heinz Henne and Arnold Michalowski. 3 vols. Heidelberg: Haug, 1993–7.

Bibliography

- *Krankenjournale D2–D4 (1801–1803)*. Commentary on the transcription by Iris von Hörsten. Stuttgart: Haug, 2004.
- *Krankenjournal D5 (1803–1806)*. Ed. Helen Varady and Arnold Micalowski. Heidelberg: Haug, 1991.
- *Krankenjournal D6 (1806–1807)*. Transcription and commentary by Johanna Bußmann. 2 vols. Heidelberg: Haug, 2002.
- *Krankenjournal D16 (1817–1818)*. Transcription and commentary by Ulrich Schuricht. 2 vols. Stuttgart: Haug, 2004.
- *Krankenjournal D 22 (1821)*. Transcription and commentary by Markus Mortsch. 2 vols. Stuttgart: Haug, 2008.
- *Krankenjournal D34 (1830)*. Transcription and commentary by Ute Fischbach-Sabel. 2 vols. Heidelberg: Haug, 1998.
- *Krankenjournal D38 (1833–1835)*. Transcription and commentary by Monika Papsch. 2 vols. Heidelberg: Haug, 2007.
- *Organon der Heilkunst. Neufassung mit Systematik und Glossar von Josef M. Schmidt*. 2nd ed. Munich: Urban & Fischer, 2006. Eng. trans., *Organon of Homeopathic Medicine*. Intro. Constantine Hering. New York: Radde, 1849.
- *Organon-Synopse: Die 6 Auflagen von 1810–1842 im Überblick*. Ed. Bernhard Luft and Matthias Wischner. Heidelberg: Haug, 2001.
- *Die Pharmakotherapie Samuel Hahnemanns in der Frühzeit der Homöopathie: Edition und Kommentar des Krankenjournals Nr. 5 (1803–1806)*. Ed. Helene Varady. Diss., Munich, 1987.
- *Reine Arzneimittellehre*. Vol. 2. 3rd ed. Dresden/Leipzig: Arnold, 1933. Unveränderter Nachdruck der Ausgabe letzter Hand. Ulm: Haug, 1955. Eng. trans., *Materia Medica Pura*. Trans. R. E. Dudgeon. Vol. 2. London: Hahnemann Publishing Society, 1881.

Hecker, August Friedrich. "S. Hahnemann *Neues Organon der rationellen Heilkunde*." *Annalen der gesammten Medicin* 2.1 (July 1810): 31–75 and 2.3 (September 1810): 193–256.

Hegel, Georg Wilhelm Friedrich. *Briefe von und an Hegel*. Ed. Johannes Hoffmeister. Vol. 2. Hamburg: Meiner, 1953.

- *Werke*. Ed. Eva Moldenhauer and Karl Markus Michel. 20 vols. Frankfurt am Main: Suhrkamp, 1970.

Heinroth, Johann Christian August. *Anti-Organon oder das Irrige der Hahnemannischen Lehre im Organon der Heilkunst*. Leipzig: Hartmann, 1825.

Herder, Johann Gottfried. *Werke*. Ed. Jürgen Brummack and Martin Bollacher. 10 vols. *Schriften zu Philosophie, Literatur, Kunst und Altertum 1774–1787*. Vol. 4. Frankfurt: Deutscher Klassiker Verlag, 1994.

Hölderlin, Friedrich. *Sämtliche Werke*. Ed. Friedrich Beißner. 6 vols. Stuttgart: Kohlhammer, 1953–62.

Hufeland, Christoph Wilhelm. "Fragmenta de viribus." *Bibliothek der practischen Heilkunde* 16 (1805): 224–5.
- "Geschichte der Gesundheit des Menschengeschlechts, nebst einer physischen Karakteristik des jetzigen Zeitalters im Vergleich zu der Vorwelt. Eine Skizze." *Journal der practischen Arzneykunde und Wundarzneykunde* 31.4 (1812): 1–35.
- *Die Homöopathie*. Berlin: Reimer, 1831.
- *Journal der practischen Arzneykunde und Wundarzneykunst* 1.1 (1795): iii–xxii.
- *Die Kunst, das menschliche Leben zu verlängern*. Stuttgart: Hippokrates, 1975. English trans., *The Art of Prolonging Life*. New York: Arno Press, 1979.
- "Mein Begriff von der Lebenskraft." *Journal der practischen Heilkunde* 6 (1798): 785–96.

Humboldt, Alexander von. *Kosmos: Entwurf einer physischen Weltbeschreibung*. 5 vols. Stuttgart: Cotta, 1845–58.

Kant, Immanuel. *Werke in sechs Bänden*. Ed. Wilhelm Weichsedel. Darmstadt: Wissenschaftliche Buchgesellschaft, 1983.

Kleist, Heinrich von. *Sämtliche Werke und Briefe*. Ed. Helmut Sembdner. 2 vols. Munich: Hanser, 1977.

Lavater, Johann Caspar. *Aussichten in die Ewigkeit in Briefen an Herrn Johann Georg Zimmermann*. 4 vols. Zurich: Orell, Geßner, 1768–78.

Marc, Carl Christian Heinrich. *Allgemeine Bemerkungen über die Gifte und ihre Wirkungen im menschlichen Körper: Nach Brownschen Systeme dargestellt*. Erlangen: Palm, 1795.

Novalis. *Werke, Tagebücher und Briefe Friedrich von Hardenbergs*. Ed. Hans-Joachim Mähl and Richard Samuel. 3 vols. Munich: Carl Hanser, 1978.

Oken, Lorenz. *Lehrbuch der Naturphilosophie*. Zurich: Schulthess, 1943.
- *Über das Universum als Fortsetzung des Sinnenssystems*. Jena: Frommann, 1808.

Paracelus. *Schriften Theophrasts von Hohenheim genannt Paracelsus*. Ed. Hans Kayser. Leipzig: Insel, 1924.

Reil, Johann Christian. *Rhapsodieen über die Anwendung der psychischen Curmethode auf Geisteszerrüttungen*. Halle: Curt, 1803. Repr. Aachen: Ariadne, 2001.
- "Von der Lebenskraft." Ed. Karl Sudhoff. Leipzig: Barth, 1910. First published in *Archiv für die Physiologie* 1 (1795): 8–162.

Richter, Jean Paul. *Zerstreute Blätter*. Vol. 2. Leipzig: n.p., 1826.

Ritter, Johann Wilhelm. *Beweis, daß ein beständiger Galvanismus den Lebensproceß in dem Thierreich begleite*. Weimar: Industrie-Comptoir, 1798.
- *Fragmente aus dem Nachlasse eines jungen Physikers*. Heidelberg: Lambert Schneider, 1969.
- *Physik als Kunst*. Munich: Lindauer, 1806.

Rousseau, Jean-Jacques. "Rêveries d'un promeneur solitare." In *Oeuvres complètes*, ed. Bernard Gagnebin and Marcel Raymond, vol. 1, 993–1099. Paris: Gallimard, 1959.

Rummel, F. "Welche Verschiedenheiten bietet die Geschichte der Homöopathie und die des Brownianismus dar?" *Archiv für homöopathische Heilkunst* 5.2 (1826): 1–18.

Schelling, Friedrich Wilhelm Joseph. *Briefe und Dokumente*. Ed. Horst Furhmans. 3 vols. Bonn: Bouvier, 1962–75.

– *Historisch-Kritische Ausgabe*. Ed. Hans Baumgartner et al. Stuttgart: Frommann-Holzboog, 1976–.

– *Schellings Werke*. Ed. Manfred Schröter. 3rd ed. 12 vols. Munich: C.H. Beck, 1927–59.

Schiller, Friedrich. *Nationalausgabe*. Ed. Julius Petersen and Gerhard Fricke. 43 vols. to date. Weimar: H. Böhlaus Nachfolger, 1943–.

Schlegel, Friedrich. *Kritische Ausgabe*. Ed. Ernst Behler and Hans Eichner. 35 vols. to date. Munich: Schöningh, 1959–.

– *Literary Notebooks, 1797–1801*. Ed. Hans Eichner. Toronto: University of Toronto Press, 1957.

Schmid, Carl Christian Erhard. *Physiologie, philosophisch betrachtet*. 3 vols. Jena: Akademische Buchhandlung, 1798–1801.

Schopenhauer, Arthur. *Über den Willen in der Natur*. Vol. 3 of *Sämtliche Werke. Kleinere Schriften*. Ed. Wolfgang von Löhneysen, 301–479. Darmstadt: Wissenschaftliche Buschgesellschaft, 1962.

Schubert, Gotthilf Heinrich. *Allgemeine Naturgeschichte*. Erlangen: Palm und Enke, 1826.

– *Ansichten von der Nachtseite der Naturwissenschaft. Nachdruck der Ausgabe von 1808*. Karben: Petra Wald, 1997.

Schwenk, Theodor. *Grundlagen der Potenzforschung*. 1954. Stuttgart: Freies Geistesleben, 1972.

Spinoza, Baruch de. *Ethics*. Ed. and trans. G.H.R. Parkinson. Oxford: Oxford University Press, 2000.

Squirrell, Richard. "Observations Addressed to the Public in General on the Cow-Pox, Shewing That It Originates in Scrophula, Commonly Called the Evil; Illustrated with Cases to Prove That It Is No Security aginst the Small-Pox." London: W. Smith & Son, 1805.

Stapf, Ernst. "Zoomagnetische Fragmente, besonders in Beziehung auf die Beurtheilung und Anwendung des Mesmerism im Geiste der homöopathische Heillehre." *Archiv für die homöopathische Heilkunst* 2.2 (1823): 1–28.

Steffens, Henrik. *Beyträge zur inneren Naturgeschichte der Erde.* Part 1. Freyberg: Verlag der Crazischen Buchhandlung, 1801.
Stieglitz, Johann. *Ueber die Homöopathie.* Hanover: Hahn, 1835.
Wedekind, Georg Christian Gottlieb von. *Prüfung des homöopathischen Systems des Herrn Hahnemanns.* Darmstadt: Leske, 1825.

Secondary Works on Homeopathy

Appell, Rainer, ed. *Homöopathie und Philosophie & Philosophie der Homöopathie.* Eisenach: Bluethenstaub, 1998.
Baschin, Marion. *Ärztliche Praxis im letzten Drittel des 19. Jahrhunderts: Der Homöopath Dr. Friedrich Paul von Bönninghausen (1828–1910).* Medizin, Gesellschaft und Geschichte 52. Stuttgart: Steiner, 2014.
– *Die Geschichte der Selbstmedikation in der Homöopathie.* Essen: KVC Verlag, 2012.
– *Wer lässt sich von einem Homöopathen behandeln? Die Patienten des Clemens Maria Franz von Bönninghausen (1785–1864). Medizin, Gesellschaft und Geschichte: Jahrbuch des Instituts für Geschichte der Medizin der Robert Bosch Stiftung.* Ed. Robert Jütte. Suppl. 37. Stuttgart: Steiner Verlag, 2010.
Baur, Jacques. *Un livre sans frontières: Histoire et métamorphoses de l'Organon de Hahnemann. L'oeuvre du fondateur de l'homéopathie à travers le temps.* Lyon: Boiron, 1991.
Bayr, Georg. *Hahnemanns Selbstversuch mit der Chinarinde im Jahre 1790: Die Konzipierung der Homöopathie.* Heidelberg: Haug, 1989.
Bleul, Gerhard, ed. *Homöopathische Fallanalyse: Von Hahnemann bis zur Gegenwart – die Methoden.* Stuttgart: Haug, 2012.
Brockmeyer, Bettina. *Selbstverständnisse: Dialoge über Körper und Gemüt im frühen 19. Jahrhundert.* Göttingen: Wallstein, 2009.
Busche, Jens. *Ein homöopathisches Patientennetzwerk im Herzogtum Anhalt-Bernburg: Die Familie von Kersten und ihr Umfeld in den Jahren 1831–1835.* Quellen und Studien zur Homöopathiegeschichte 11. Stuttgart: Haug, 2008.
Czech, Barbara. *Konstitution und Typologie in der Homöopathie des 19. und 20. Jahrhunderts.* Heidelberg: Haug, 1996.
Dinges, Martin, "Arztpraxen 1500–1900. Zum Stand der Forschung." In *Arztpraxen im Vergleich: 18.–20. Jahrhundert,* ed. Elisabeth Dietrich-Daum, Martin Dinges, Robert Jütte, and Christine Roilo, 23–61. Innsbruck: StudienVerlag, 2008.
– "Bettine von Arnim (1785–1859), eine für die Homöopathie engagierte Patientin. Handlungsräume in Familie, Landgut und öffentlichem Raum/

Politik." *Orvostörténeti Közlemények / Communicationes de historia artis medicinae* 49.1–2 (2004): 105–22.
- "Hahnemanns Falldokumentation in historischer Perspektive." *Naturheilpraxis* 11 (2010): 1356–62.
- "Men's Bodies 'Explained' on a Daily Basis in Letters from Patients to Samuel Hahnemann (1830–35). In *Patients in the History of Homeopathy*, ed. Martin Dinges, 85–118. Sheffield: European Association for the History of Medicine and Health Publications, 2002.

Dinges, Martin, ed. *Homöopathie: Patienten, Heilkundige, Institutionen: Von den Anfängen bis heute*. Heidelberg: Haug, 1996.
- *Patients in the History of Homeopathy*. Sheffield: European Association for the History of Medicine and Health Publications, 2002.

Dinges, Martin, and Klaus Holzapfel. "Von Fall zu Fall: Falldokumentation und Fallredaktion Clemens von Bönninghausen und Annette von Droste-Hülshoff." *Zeitschrift für klassische Homöopathie* 48.4 (2004): 149–67.

Eppenich, Heinz. "Samuel Hahnemann und die Beziehung zwischen Homöopathie und Mesmerismus." *Zeitschrift für klassische Homöopathie* 38.4 (1994): 153–60.

Faure, Oliver. "Behandlungsverläufe. Die französischen Patienten von Samuel und Mélanie Hahnemann (1834–1868)." In *Medizin, Gesellschaft und Geschichte: Jahrbuch des Instituts für Geschichte der Medizin der Robert Bosch Stiftung*, ed. Martin Dinges and Vincent Barras, 197–210. Suppl. 29. Stuttgart: Franz Steiner, 2007.

Fink, Gottfried Wilhelm. *Kurze Geschichte der Homöopathie: Entdeckung und Entfaltung der sanften Heilkunde und Medizin*. Leipzig: Bohmeier, 2009.

Fräntzki, Ekkehard. *Die Idee der Wissenschaft bei Samuel Hahnemann*. Heidelberg: Haug, 1976.

Gantenbein, Urs Leo. "Der Einfluß der Ersten Wiener Schule auf das Arnzeiverständnis bei Samuel Hahnemann." *Medizin, Gesellschaft und Geschichte* 19 (2000): 229–49.
- "Similia Similibus: Samuel Hahnemann und sein Schatten Paracelsus." *Nova Acta Paracelsica* 13 (1990): 293–328.

Gawlik, Willibald. *Samuel Hahnemann: Synchronopse seines Lebens: Geschichte, Kunst, Kultur und Wissenschaft bei Entstehung der Homöopathie 1755–1843*. Stuttgart: Sonntag, 1996.

Genneper, Thomas. *Als Patient bei Samuel Hahnemann: die Behandlung Friedrich Wiecks in den Jahren 1815/1816*. Heidelberg: Haug, 1990.

Große-Onnebrink, Jörg. *Der Gottesbegriff bei Samuel Hahnemann*. Diss., Münster, 2004.

Haehl, Richard. *Samuel Hahnemann: His Life and Work*. Trans. Marie L. Wheeler and W. Grundy. 2 vols. London: London Homeopathic Publ. Co., 1922.

Handley, Rima. *Auf den Spuren des späten Hahnemann*. Stuttgart: Sonntag, 2001.

– *In Search of the Later Hahnemann*. Buckinghamshire, UK: Beaconsfield, 1997.

Heinz, Inge Christine. *"Schicken Sie Mittel, senden Sie Rath!" Prinzessin Luise von Preußen als Patientin Samuel Hahnemanns in den Jahren 1829–1835*. Quellen und Studien zur Homöopathiegeschichte 15. Essen : KVC Verlag, 2011.

Heinz, Inge Christine, and Matthias Wischner. "Hahnemann und die Pockenimpfung." *Zeitschrift für klassische Homöopathie* 56.4 (2012): 180–8.

Heinze, Sigrid, ed. *Homöopathie 1796–1996: Eine Heilkunde und ihre Geschichte*. Berlin: Edition Lit.europe, 1996.

Hess, Volker. "Samuel Hahnemann und die Semiotik." *Medizin, Gesellschaft und Geschichte* 12 (1993): 177–204.

– "Zeichen ohne Differenz: Hahnemanns Reform des semiotischen Aufschreibesystems." In *Homöopathie und Philosophie & Philosophie und Homöopathie*, ed. Rainer G. Appell, 67–86. Eisenach: Bluethenstaub, 1998.

Hickmann, Reinhard. *Das Psorische Leiden der Antonie Volkmann. Edition und Kommentar einer Krankengeschichte aus Hahnemanns Krankenjournalen von 1829–1831*. Quellen und Studien zur Homöopathiegeschichte 2. Heidelberg: Haug, 1996.

Just, Claus. *Der Akt der Ähnlichkeit: Wissenschaft. Therapie. Kunst*. Heidelberg: Haug, 1994.

Jütte, Robert. "Die Arzt-Patient-Beziehung im Spiegel der Krankenjournale Samuel Hahnemanns." In *Arztpraxen im Vergleich: 18.–20. Jahrhundert*, ed. Elisabeth Dietrich-Daum, Martin Dinges, Robert Jütte, and Christine Roilo, 109–27. Innsbruck: StudienVerlag, 2008.

– "Case Taking in Homeopathy in the 19th and 20th Centuries." *British Homeopathic Journal* 87 (1998): 39–47.

– *The LM Potencies in Homeopathy: From Their Beginning to the Present Day*. Stuttgart: Institut für Geschichte der Medizin, 2008.

– *Samuel Hahnemann: Begründer der Homöopathie*. Munich: dtv, 2005. English translation: http://www.igm-bosch.de/content/language1/downloads/SamuelHahnemannTheFounderofHomeopathy.pdf.

– "200 Jahre Simile-Prinzip: Magie-Medizin-Metaphor." *Allgemeine Homöopathische Zeitung* 242 (1997): 3–16.

Klunker, Will. "Hahnemanns historische Begründung der Psoralehre." *Zeitschrift für klassische Homöopathie* 34 (1990): 3–13.

Lucae, Christian. "Hahnemanns Prüfungsprotokolle." *Zeitschrift für klassische Homöopathie* 56.12 (2012): 4–17.

Lucae, Christian, and Gunnar Stollberg, et al. Special issue, "Zur Geschichte der Homöopathie und alternativer Heilweisen." *Medizin, Gesellschaft und Geschichte* 18 (1999): 81–208.

Lucae, Christian, and Matthias Wischner. "Rein oder nicht rein? Zur Quellenlage von Hahnemanns Arzneimittellehre." *Zeitschrift für klassische Homöopathie* 54.1 (2010): 13–22.

Mayr, Stefan. *Herstellung homöopathischer Arzneimittel – von Hahnemann bis zu Schwabes Pharmakopöe 1872*. Quellen und Studien zur Homöopathiegeschichte 20. Essen: KVC, 2014.

Minder, Peter. *Gesamtregister zu Hahnemanns Werk: Sach-, Arznei- und Personenverzeichnis*. Stuttgart: Haug, 2002.

Müller, Carl Werner. *Gleiches zu Gleichem: Ein Prinzip frühgriechischen Denkens*. Wiesbaden: Otto Harrassowitz, 1965.

Mure, Corine. "Stoerck, Cullen, Hahnemann: Aux sources de la similitude." *Le Journal de l'homéopathie*, supplement 8 (May 1996): 30–43.

Papsch, Monika. "Sozialstatistische Auswertung von Samuel Hahnemanns (1755–1843) homöopathischer Praxis von Dezember 1833 bis Mai 1835 anhand seines Krankentagebuches 'D38.'" In *Arztpraxen im Vergleich: 18.–20. Jahrhundert*, ed. Elisabeth Dietrich-Daum, Martin Dinges, Robert Jütte, and Christine Roilo, 129–45. Innsbruck: StudienVerlag, 2008.

Plate, Uwe. *Hahnemanns Arbeitsweise mit den Symptomen: Dargestellt an Praxisfällen aus seinen Krankenjournalen*. Braunschweig: Lexikon, 2003.

Sauerbeck, K.O. "Wie gelangte Hahnemann zu den hohen Potenzen? Ein Kapitel aus der Geschichte der Homöopathie." *Allgemeine Homöopathische Zeitung* 235 (1990): 223–32.

Schmidt, Josef M. *Bibliographie der Schriften Samuel Hahnemanns*. Rauenberg: Siegle,1989.

– "Die Entstehung, Verbreitung und Entwicklung von Heilsystemen als Gegenstand der Medizingeschichte – am Beispiel der Homöopathie." *Sudhoffs Archiv* 91.1 (2007): 38–72.

– "Homöopathie und Philosophie." *Scheidewege: Jahresschrift für skeptisches Denken* 20 (1990–1): 141–65.

– "Die literarischen Belege Samuel Hahnemanns für das Simile-Prinzip." *Jahrbuch des Instituts für Geschichte der Medizin* 7 (1988): 161–87.

– *Die philosophischen Vorstellungen Samuel Hahnemanns bei der Begründung der Homöopathie*. Munich: Sonntag, 1990.

– "Die Publikationen Samuel Hahnemanns." *Sudhoffs Archiv* 72 (1988): 14–36.

- "Samuel Hahnemann und das Ähnlichkeitsprinzip." *Medizin, Gesellschaft und Geschichte* 29 (2010): 151–84.
Schott, Heinz. "Die Bedeutung des ärztlichen Selbstversuchs in der Medizingeschichte." In *Der verwundete Heiler: Homöopathie und Psychoanalyse im Gespräch*, ed. Rainer G. Appell, 13–33. Heidelberg: Haug, 1995.
Schreiber, Kathrin. *Samuel Hahnemann in Leipzig: Die Entwicklung der Homöopathie zwischen 1811 und 1821: Förderer, Gegner und Patienten*. Quellen und Studien zur Homöopathiegeschichte 8. Stuttgart: Haug, 2002.
Schultz, Hartwig. "Kunst und Homöopathie: Unbekannte Briefzeugnisse aus Bettine von Arnims Korrespondenz mit Karl Friedrich und Susanne Schinkel." *Internationales Jahrbuch der Bettina-von-Arnim-Gesellschaft: Forum für die Erforschung von Romantik und Vormärz* 20/21 (2009): 37–56.
Schwanitz, Hans Joachim. *Homöopathie und Brownianismus 1795–1844: Zwei wissenschaftstheoretische Fallstudien aus der praktischen Medizin*. Stuttgart: Gustav Fischer, 1983.
Schriewer, Miriam Leoni. "'Kann der Körper genesen, wo die Seele so gewaltig krankt?' Weibliche Gemüts- und Nervenleiden in der Patientenkorrespondenz Hahnemanns am Beispiel der Kantorstochter Friederike Lutze (1798–1878)." Diss., Würzburg, 2011.
Seiler, Hanspeter. *Die Entwicklung von Samuel Hahnemanns ärztlicher Praxis*. Heidelberg: Haug, 1988.
Sloterdijk, Peter. "Die Andersheilenden: Über einige Aspekte der Homöopathie im Lichte des philosophischen Therapiegedankens der Neuzeit." In *200 Jahre Homöopathie: Festakt in der Paulskirche zu Frankfurt am Main. 14. September 1996*, 9–38. Heidelberg: Haug, 1996.
Thoms, Ulrike. "Konfliktfall Homöopathie: Die klinischen Versuche zur Prüfung des Wertes der Homöopathie beim Militär und in der Berliner Charité 1820 bis 1840." *Medizin, Gesellschaft und Geschichte* 21 (2002): 173–218.
Tischner, Rudolf. *Geschichte der Homöopathie*. Leipzig: Schwabe, 1939.
- "Goethe und das Biologische Grundgesetz." *Allgemeine Homöopathische Zeitung* 184.6 (1936): 402–10.
- "Hahnemann und Goethe: Ein Vergleich." *Hippokrates: Zeitschrift für praktische Heilkunde* 18 (1947): 265–78.
- "Über den Begriff des Dynamischen bei Hahnemann." *Allgemeine Homöopathische Zeitung* 180.4/5 (1932): 358–67.
Treuherz, Francis. *Genius of Homeopathy*. Glasgow: Saltire Books, 2010.
Vieracker, Viktoria. *Nosoden und Sarkoden: Einführung und Entwicklzng zweier homöopathischer Arzneimittelgruppen in der ersten Hälfte des 19. Jahrhunderts*. Quellen und Studien zur Homöopathiegeschichte 18. Essen: KVC, 2013.

Walach, Harald. "Homöopathie und die moderne Semiotik." *Jahrbuch des Instituts für Geschichte der Medizin* 7 (1988): 135–60.
- "Magic of Signs: A Non-Local Interpretation of Homeopathy." *British Homeopathic Journal* 89 (2000): 127–40.
- "Methoden der homöopathischen Arzneimittelprüfung. Teil 1: Historische Entwicklung und Stand der Forschung." In *Naturheilverfahren und unkonventionelle medizinische Richtungen*, ed. M. Bühring and F. H. Kemper, 1–42. Berlin: Springer, 1999.

Wischner, Matthias. "Die Benutzung von Repertorien in Hahnemanns Pariser Praxis. Teil 1: Die verwendeten Repertorien." *Zeitschrift für klassische Homöopathie* 56.2 (2012): 60–73.
- "Die Benutzung von Repertorien in Hahnemanns Pariser Praxis. Teil 2: Der Einfluss auf die Arzneimittelwahl." *Zeitschrift für klassische Homöopathie* 56.3 (2012): 127–36.
- *Fortschritt oder Sackgasse? Die Konzeption der Homöopathie in Samuel Hahnemanns Spätwerk (1824–1842)*. Essen : KVC Verlag, 2000.
- *Kleine Geschichte der Homöopathie*. Essen: KVC Verlag, 2004.

Wittern, Renate. "Zum Verhältnis von Homöopathie und Mesmerismus." In *Franz Anton Mesmer und die Geschichte des Mesmerismus*, ed. Heinz Schott, 108–15. Stuttgart: Steiner, 1985.
- "Zur Geschichte der Homöopathie und alternativer Heilweisen." *Medizin, Gesellschaft und Geschichte* 18 (1999).

Secondary Works on History of Science, Medicine, Philosophy, and Literature

Aesch, Alexander Gode-von. *Natural Science in German Romanticism*. New York: AMS Press, 1966.

Agamben, Giorgio. "What Is a Paradigm?" In *The Signature of All Things: On Method*, 9–32. New York: Zone, 2009.

Amrine, Frederick. "Goethean Intuitions." *Goethe Yearbook* 18 (2011): 35–50.
- "'The Magic Formula We All Seek': Spinoza + Fichte = x." In *Religion, Reason, and Culture in the Age of Goethe*, ed. Elisabeth Krimmer and Patricia Anne Simpson, 244–65. Rochester, NY: Camden House, 2013.

Arz, Maike. *Literatur und Lebenskraft: Vitalistische Naturforschung und bürgerliche Literatur um 1800*. Stuttgart: Verlag für Wissenschaft und Forschung, 1996.

Auf der Horst, Christoph. "Vorstellungen, Ideen, Begriffe: Intellectual History in der Medizingeschichtsschreibung am Beispiel des Naturbegriffs." In *Medizingeschichte: Aufgaben, Probleme, Perspektiven*, ed. Norbert Paul and Thomas Schlich, 186–215. Frankfurt: Campus, 1998.

Bark, Irene. "'Spur der Empfindung im anorganischen Reich': Novalis' Poetik des Galvanismus im Kontext der frühromantischen Philosophie und Naturwissenschaft." In *Faktenglaube und fiktionales Wissen: Zum Verhältnis von Wissenschaft und Kunst in der Moderne*, ed. Daniel Fulda and Thomas Prüfer, 93–125. Kölner Studien zur Literaturwissenschft 9. Frankfurt: Lang, 1996.

Barkhoff, Jürgen. *Magnetische Fiktionen: Literarisierung des Mesmerismus in der Romantik*. Stuttgart: Metzler, 1995.

– "Romantic Science and Psychology." In *The Cambridge Companion to German Romanticism*, ed. Nicholas Saul, 209–25. Cambridge: Cambridge University Press, 2009.

Bell, Matthew. *The German Tradition of Psychology in Literature and Thought, 1700–1840*. Cambridge: Cambridge University Press, 2005.

Bennett, Jane. *Vibrant Matter: A Political Ecology of Things*. Durham, NC: Duke University Press, 2010.

Bergengruen, Maximilian. "Magischer Organismus: Ritters und Novalis' 'Kunst, die Natur zu modificiren.'" In *Ästhetische Erfindung der Moderne? Perspektiven und Modelle 1750–1850*, ed. Britta Herrmann and Barbara Thums, 39–54. Würzburg: Könighausen & Neumann, 2003.

– *Nachfolge Christi – Nachahmung der Natur: Himmlische und natürliche Magie bei Paracelsus, im Paracelsismus und in der Barockliteratur (Scheffler, Zesen, Grimmelshausen)*. Hamburg: Meiner, 2007.

– "Signatur, Hieroglyphe, Wechselrepräsentation. Zur Metaphysik der Schrift in Novalis' *Lehrlingen*." *Athenäum: Jahrbuch für Romantik* 14 (2004): 43–67.

Bertalanffy, Ludwig von. *General Systems Theory: Foundations, Development, Application*. Chicago: University of Illinois Press, 1963.

Botsch, Walter. *Die Bedeutung des Begriffs Lebenskraft für die Chemie zwischen 1750 und 1850*. Diss., Stuttgart, 1997.

Böhme, Hartmut. "Denn nichts ist ohne Zeichen. Die Sprache der Natur: Unwiederbringlich?"; "Lebendige Natur: Wissenschaftskritik, Naturforschungund allegorische Hermetik bei Goethe"; "Der sprechende Leib. Die Semiotiken des Körpers am Ende des 18. Jahrhunderts und ihre hermetische Tradition." In *Natur und Subjekt*, 38–66, 145–78, 179–211. Frankfurt: Suhrkamp, 1988.

Borgards, Roland. *Poetik des Schmerzes: Physiologie und Literatur von Brockes bis Büchner*. Munich: Fink, 2007.

Brandstetter, Gabrielle, Gerhard Neumann, and Alexander von Bormann, eds. *Romantisches Wissenspoetik: Die Künste und die Wissenschaften um 1800*. Würzburg: Königshausen & Neumann, 2004.

Breithaupt, Fritz. "The Invention of Trauma in German Romanticism." *Critical Inquiry* 32.1 (Autumn, 2005): 77–101.

Broman, Thomas Hoyt. *The Transformation of German Academic Medicine, 1750–1820.* Cambridge: Cambridge University Press, 1996.

Brown, Jane K. *Goethe's Allegories of Identity.* Philadelphia: University of Pennsylvania Press, 2014.

Buchinger, Helene. "Arztgestalten bei Goethes *Faust.*" Diss., Technical University Munich, 2012.

Chaouli, Michel. *The Laboratory of Poetry: Chemistry and Poetics in the Works of Friedrich Schlegel.* Baltimore: Johns Hopkins University Press, 2002.

Coulter, Harris. *Divided Legacy: A History of the Schism in Medical Thought.* Vol. 2, *Progress and Regress: J.B. Van Helmont to Claude Bernard.* Washington, DC: Wehawken, 1977.

Cunningham, Andrew. "Medicine to Calm the Mind: Boerhaave's Medical System, and Why It Was Adopted in Edinburgh." In *The Medical Enlightenment of the Eighteenth Century*, ed. Andrew Cunningham and Roger French, 40–66. Cambridge: Cambridge University Press, 1990.

Cunningham, Andrew, and Nicholas Jardine, eds. *Romanticism and the Sciences.* Cambridge: University Press of Cambridge, 1990.

Daiber, Jürgen. *Experimentalphysik des Geistes: Novalis und das romantische Experiment.* Göttingen: Vandenhoeck & Ruprecht, 2001.

– "'Experimentieren mit dem Tode': Zur spezifischen Form und Praxis eines romantischen Selbstexperiments." In *Literarische Experimentalkulturen: Poetologien des Experiments im 19. Jahrhundert*, ed. Marcus Krause and Nicolas Pethes, 103–21. Würzburg: Königshausen & Neumann, 2005.

– "Selbstexperimentation: Von der empirisch-aufklärerischen zu einer spezifisch romantischen Versuchspraxis." *Aurora* 58 (1998): 49–68.

Darnton, Robert. *Mesmerism and the End of Enlightenment in France.* Cambridge, MA: Harvard University Press, 1968.

Daston, Lorraine, and Peter Galison. *Objectivity.* Cambridge, MA: Zone, 2007.

Deleuze, Gilles. *Expressionism in Philosophy: Spinoza.* Trans. Martin Joughin. New York: Zone, 1992.

– *Spinoza: Practical Philosophy.* Trans. Robert Hurley. San Francisco: City Lights Books, 1988.

Deleuze, Gilles, and Félix Guattari. *A Thousand Plateaus: Capitalism and Schizophrenia.* Trans. Brian Massumi. Minneapolis: University of Minnesota Press, 1987.

– *What Is Philosophy?* New York: Columbia University Press, 1994.

DeLong, Anne. *Mesmerism, Medusa, and the Muse: The Romantic Discourse of Spontaneous Creativity.* Lanham, MD: Lexington Books, 2012.

Dickson, Sheila, Stefan Goldmann, and Christof Wingertszahn, eds. *"Fakta, und kein moralisches Geschwätz": Zu den Fallgeschichten im*

Magazin zur "Erfahrungsseelenkunde" (1783–1793). Göttingen: Wallstein, 2011.

Dinges, Martin. "Medizinische Aufklärung bei Johann Georg Zimmermann: Zum Verhältnis von Macht und Wissen bei einem Arzt der Aufklärung." In *Schweizer im Berlin des 18. Jahrhunderts*, ed. Martin Fontius and Helmut Holzhey, 137–50. Berlin: Akademie, 1996.

Dixon, Thomas. *From Passions to Emotions: The Creation of a Secular Psychological Category*. Cambridge: Cambridge University Press, 2003.

Dreißigacker, Erdmuth. *Populärmedizinische Zeitschriften des 18. Jahrhunderts zur hygienischen Volksaufklärung*. Diss., Marburg, 1970.

Duden, Barbara. *Geschichte unter der Haut: Ein Eisenacher Arzt und seine Patientinnen um 1730*. Stuttgart: Klett-Cotta, 1987.

Düwell, Susanne, and Nicolas Pethes, eds. *Fall–Fallgeschichte–Fallstudies: Theorie und Geschichte einer Wissensform*. Frankfurt: Campus, 2014.

Dyck, Martin. *Novalis and Mathematics: A Study of Friedrich von Hardenberg's Fragments on Mathematics and Its Relation to Magic, Music, Religion, Philosophy, Language, and Literature*. Chapel Hill: University of North Carolina Press, 1960.

Eckart, Wolfgang Uwe. "'Und setzet eure Worte nicht auf Schrauben': Medizinische Semiotik vom Ende des 18. bis zum Beginn des 20. Jahrhunderts – Gegenstand und Forschung." *Berichte zur Wissenschaftsgeschichte* 19 (1996): 1–18.

Eckart, Wolfgang Uwe, and Robert Jütte. *Medizingeschichte: Eine Einführung*. Cologne: Böhlau, 2007.

Egger, Irmgard. *Diätetik und Askese: Zur Dialektik der Aufklärung in Goethes Romanen*. Munich: Fink, 2001.

Engelhardt, Dietrich von. "Natural Science in the Age of Romanticism." In *Modern Esoteric Spirituality*, ed. Antoine Faivre and Jacob Needleman, 101–31. New York: Crossroads, 1992.

– "Naturforschung im Zeitalter der Romantik." In *"Fessellos durch die Systeme": Frühromantisches Naturdenken im Umfeld von Arnim, Ritter und Schelling*, ed. Walther Ch. Zimmerli, Klaus Stein, and Michael Gerten, 19–48. Stuttgart–Bad Cannstatt: Frommann-Holzboog, 1997.

– "Naturwissenschaft zwischen Empirie und Metaphysik um 1830: Unter besonderer Berücksichtigung der Entwicklung in Österreich." In *Verdrängter Humanismus, Verzögerte Aufklärung*, ed. Michael Benedikt and Reinhold Knoll, vol. 3, 399–407. Vienna: Triade, 1995.

– "Novalis im medizinhistorischen Kontext." In *Novalis und die Wissenschaften*, ed. Herbert Uerlings, 65–85. Tübingen: Max Niemeyer, 1997.

- "Schellings philosophische Grundlegung der Medizin." In *Natur und geschichtlicher Progress: Studien zur Naturphilosophie F. W. J. Schellings*, ed. Hans-Jörg Sandkühler, 305–25. Frankfurt: Suhrkamp, 1984.
Engels, Eve-Marie. "Lebenskraft." In *Historisches Wörterbuch der Philosophie*, ed. J. Ritter and K. Gründer, vol. 5, 122–8. Darmstadt: Wissenschaftliche Buchgesellschaft, 1980.
Erpenbeck, John. "'. . . die Gegenstände der Natur an sich selbst . . .': Subjekt und Objekt in Goethes naturwissenschaftlichem Denken seit der italienischen Reise." *Goethe Jahrbuch* 105 (1988): 212–33.
Fischer, Hans. "Die Krankheitsauffassung Friedrich von Hardenbergs (Novalis): Poesie und Kunst als Erkenntnisgrundlage der romantischen Medizin." *Verhandlungen der Naturforschenden Gesellschaft Basel* 56 (1945): 390–410.
Fletcher, Angus. *Allegory: The Theory of a Symbolic Mode*. Ithaca: Cornell University Press, 1964.
Förster, Eckhart. *The Twenty-five Years of Philosophy: A Systematic Reconstruction*. Cambridge, MA: Harvard University Press, 2012.
Foucault, Michel. *The Birth of the Clinic: An Archaeology of Medical Perception*. Trans. A.M. Sheridan. London: Routledge, 1989.
- *The Order of Things: An Archaeology of the Human Sciences*. New York: Vintage, 1973.
Frank, Manfred. *Selbstgefühl: Eine historisch-systematische Erkundung*. Frankfurt: Suhrkamp, 2002.
Frazer, James George. *The Golden Bough: A Study in Magic and Religion*. Abridged edition. New York: Macmillan, 1956.
French, Roger. "Sickness and the Soul: Stahl, Hoffmann and Sauvages on Pathology." In *The Medical Enlightenment of the Eighteenth Century*, ed. Andrew Cunningham and Roger French, 88–110. Cambridge: Cambridge University Press, 1990.
Frevert, Ute, et al. *Gefühlswissen: Eine lexikalische Spurensuche in der Moderne*. Frankfurt: Campus, 2011.
Fromm, Waldemar. "Inspirierte Ähnlichkeit: Überlegungen zu einem poetischen Verfahren des Novalis." *Deutsche Vierteljahrsschrift* 71 (1977): 559–88.
- "Die Sympathie des Zeichens mit dem Bezeichneten: Ähnlichkeit in Literatur und Sprachästhetik um 1800." *Ästhetik des Ähnlichen*, ed. Gerald Funk, 35–67. Frankfurt: Fischer, 2001.
Gafner, Lina, Iris Ritzmann, and Katharina Weikl, eds. "Penning Patient's Histories: Doctors' Records and the Medical Market in the 18th and 19th Century." Special issue of *Gesnerus* 69.1 (2012).

Gaier, Ulrich. "Naturzeichen: Von Paracelsus bis Novalis." In *Die Unvermeidlichkeit der Bilder*, ed. Gerhard von Graevenitz, 117–32. Tübingen: Narr, 2001.

Gerabek, Werner E. *Friedrich Wilhelm Joseph Schelling und die Medizin der Romantik: Studien zu Schellings Würzburger Periode*. Frankfurt am Main: Lang, 1995.

Geyer-Kordesch, Johanna. "Georg Ernst Stahl's Radical Pietist Medicine and Its Influence on the German Enlightenment." In *The Medical Enlightenment of the Eighteenth Century*, ed. Andrew Cunningham and Roger French, 67–87. Cambridge: Cambridge University Press, 1990.

— "Medizinische Fallbeschreibungen und ihre Bedeutung in der Wissenschaftsreform des 17. und 18. Jahrhunderts." *Jahrbuch des Instituts für Geschichte der Medizin der Robert Bosch Stiftung* 9 (1990): 7–19.

Gigante, Denise. *Life: Organic Form and Romanticism*. New Haven: Yale University Press, 2009.

Goldmann, Stefan. *Christoph Wilhelm Hufeland im Goethekreis: Eine psychoanalytische Studie zur Autobiographie und ihre Topik*. Stuttgart: M&P, 1993.

— "Von der Lebenskraft zum Unbewussten: Konzeptwandel in der Anthropologie um 1800." In *Homöopathie und Philosophie & Philosophie und Homöopathie*, ed. Rainer G. Appell, 149–74. Eisenach: Bluethenstaub, 1998.

Goldstein, Jan. *Console and Classify: The French Psychiatric Profession in the Nineteenth Century*. Cambridge: Cambridge University Press, 1987.

— *The Post-Revolutionary Self: Politics and Psyche in France, 1750 to 1850*. Cambridge, MA: Harvard University Press, 2005.

Golinski, Jan. "Humphry Davy: The Experimental Self." *18th Century Studies* 45.1 (Fall 2011): 14–28.

Goodbody, Axel. *Natursprache: Ein dicthtungstheoretisches Konzept der Romantik und seine Wiederaufnahme in der modernen Naturlyrik (Novalis–Eichendorff–Lehmann–Eich)*. Neumünster: Wachholtz, 1984.

Gray, Richard. *About Face: German Physiognomic Thought from Lavater to Auschwitz*. Detroit: Wayne State University Press, 2004.

Hadot, Pierre. *The Veil of Isis: An Essay on the History of the Idea of Nature*. Trans. Michael Chase. Cambridge, MA: University Press of Harvard, 2006.

Haller, John S. *The History of American Homeopathy from Rational Medicine to Holisitic Health Care*. New Brunswick, NJ: Rutgers University Press, 2009.

Heinz, Jutta. *Wissen vom Menschen und Erzählen vom Einzelfall: Untersuchungen zum anthropologischen Roman der Spätaufklärung*. Berlin: de Gruyter, 1996.

Henderson, Fergus. "Novalis, Ritter and 'Experiment': A Tradition of 'Active Empiricism.'" In *The Third Culture: Literature and Science*, ed. Eleanor S. Shaffer, 153–69. Berlin: de Gruyter, 1998.

Henne, Heinz. "Probleme um die ärztliche Diagnose als Grundlage für die Therapie zu Ende des 18. und in der ersten Hälfte des 19. Jahrhunderts." In *Medizinische Diagnostik in der Geschichte und Gegenwart: Festschrift für Heinz Goerke*, ed. Christa Habrich, Frank Marguth and Jörg Henning Wolf, 283–96. Munich: Fritsch, 1978.

Hess, Volker. "Diagnose und Krankheitsverständnis der medizinischen Klinik der Berliner Universität zwischen 1820 und 1845." In *Die Medizin an der Berliner Universität und an der Charité zwischen 1810 und 1850*, ed. Peter Schenk and Hans-Uwe Lammel, 101–10. Abhandlungen zur Geschichte der Medizin und der Naturwissenschaft 67. Matthiesen: Husum, 1995.

Hindrichs, Gunnar. "Goethe's Notion of an Intuitive Power of Judgment." *Goethe Yearbook* 18 (2011): 51–65.

Hitzer, Bettina. "Gefühle heilen." In *Gefühlswissen: Eine lexikalische Spurensuche in der Moderne*, ed. Ute Frevert et al., 121–51. Frankfurt: Campus, 2011.

Holland, Jocelyn. *German Romanticism and Science: The Procreative Poetics of Goethe, Novalis, and Ritter*. New York: Routledge, 2009.

Holmes, Richard. *The Age of Wonder: How the Romantic Generation Discovered the Beauty and Terror of Science*. London: HarperCollins, 2008.

Irmscher, Hans Dietrich. *"Weitstrahlsinniges" Denken: Studien zu Johann Gottfried Herder*. Ed. Marion Heinz and Violetta Stolz. Würzburg: Königshausen & Neumann, 2009.

Jahne, Hans Niels. "Mathematik und Romantik." In *Diszplinen im Kontext: Perspektiven der Disziplingeschichtsschreibung*, ed. Volker Peckhaus and Christian Thiel, 163–98. Munich: Fink, 1999.

Jantzen, Jörg. "Physiologische Theorien." In Friedrich Wilhelm Schelling, *Historisch-Kritische Ausgabe*, Reihe 1, *Ergänzungsband zu Werke Band 5 bis 9: Wissenschaftshistorischer Bericht zu Schellings Naturphilosophischen Schriften 1797–1800*, 373–668. Stuttgart: Frommann-Holzboog, 1994.

Janz, Rolf-Peter, Fabians Stoermer, and Andreas Hiepko, eds. *Schwindelerfahrungen: Zur kulturhistorischen Diagnose eines vieldeutigen Symptoms*. Amsterdam: Rodopi, 2003.

Jay, Martin. *Songs of Experience: Modern American and European Variations on a Universal Theme*. Berkeley: University of California Press, 2005.

Jewson, Nicolas D. "Disappearance of the Sick-Man from Medical Cosmology, 1770–1870." *Sociology: The Journal of the British Sociological Association* 10 (1976): 225–44.

John, David. "The Duality of Goethe's Materialism." *Lumen. Selected Proceedings from the Canadian Society for Eighteenth-Century Studies* 32 (2013): 57–71.

Kaufmann, Doris. "Schmerz zur Heilung des Selbst: Heroische Kuren in der Psychiatrie des frühen 19. Jahrhunderts." *Medizin, Gesellschaft und Geschichte* 15 (1996): 101–16.

Kelley, Theresa M. *Clandestine Marriage: Botany and Romantic Culture.* Baltimore: Johns Hopkins University Press, 2012.

Koschorke, Albrecht. *Körperströme und Schriftverkehr: Mediologie des 18. Jahrhunderts.* Munich: Fink, 1999.

– "Poiesis des Leibes: Johann Christian Reils romantische Medizin." In *Romantische Wissenspoetik: Die Künste und die Wissenschaften um 1800*, ed. Gabriele Brandstetter and Gerhard Neumann, 259–72. Würzburg: Königshausen & Neumann, 2004.

Krell, David. *Contagion: Sexuality, Disease, and Death in German Idealism and Nationalism.* Indianapolis: University Press of Indiana, 1998.

Kroker, Kenton, Jennifer Keelan, and Pauline M.H. Mazumdar, eds. *Crafting Immunity: Working Histories of Clinical Immunology.* Burlington, VT: Ashgate, 2008.

Kuhn, Bernhard. *Autobiography and Natural Science in the Age of Romanticism: Rousseau, Goethe, Thoreau.* Burlington, VT: Ashgate, 2009.

Kutzer, Michael. "'Psychiker' als 'Somatiker' – 'Somantiker' als 'Psychiker': Zur Frage der Gültigkeit psychiatriehistorischer Kategorien." *Medizinhistorisches Journal* 38.1 (2003): 17–33.

– "Stimulation, Regulation, Repression: Reiz als zentraler Begriff der frühen Psychiatrie." In *Reiz Imagination Aufmerksamkeit: Erregung und Steuerung von Einbildungskraft im klassischen Zeitalter (1680–1830)*, ed. Jörn Steigerwald and Daniela Watzke, 270–9. Würzburg: Königshausen & Neumann, 2003.

Kuzniar, Alice A. *Delayed Endings: Nonclosure in Novalis and Hölderlin.* Athens: University of Georgia Press, 1987.

– "Hearing Woman's Voices in *Heinrich von Ofterdingen*." *PMLA* 107 (1992): 1196–1207.

– "A Higher Language: Novalis on Communion with Animals." *German Quarterly* 76 (2003): 426–42.

Labisch, Alfons, and Reinhard Spree, eds. *"Einem jeden Kranken in einem Hospitale sein eigenes Bett": Zur Sozialgeschichte des Allgemeinen Krankenhauses in Deutschland im 19. Jahrhundert.* Frankfurt: Campus, 1996.

Lachmund, Jens, and Gunnar Stollberg, eds. *Patientenwelten: Krankheit und Medizin vom späten 18. bis zum frühen 20. Jahrhundert im Spiegel von Autobiographien.* Opladen: Leske & Budrich, 1996.

- *The Social Construction of Illness*. Medizin, Gesellschaft und Geschichte: Jahrbuch des Instituts für Geschichte der Medizin der Robert Bosch Stiftung. Suppl. 1. Stuttgart: Franz Steiner, 1992.
Lacoue-Labarthe, Philippe, and Jean-Luc Nancy. *The Literary Absolute: The Theory of Literature in German Romanticism*. Trans. Philip Barnard and Cheryl Lester. Albany: SUNY Press, 1988.
Lange, Horst. "Goethe and Spinoza: A Reconsideration." *Goethe Yearbook* 18 (2011): 11–33.
Leibbrand, Werner. *Die speculative Medizin der Romantik*. Hamburg: Claassen, 1956.
Lenoir, Timothy. *The Strategy of Life: Teleology and Mechanics in 19th-Century Biology*. Dordrecht: Reidel, 1982.
Levin, Karl-Heinz. "Krankheiten – historische Deutung versus retrospective Diagnose." In *Medizingeschichte: Aufgaben, Probleme, Perspektiven*, ed. Norbert Paul and Thomas Schlich, 153–85. Frankfurt: Campus, 1998.
Lindemann, Mary. *Health and Healing in Eighteenth-Century Germany*. Baltimore: Johns Hopkins University Press, 1996.
Lobo, Francis M. "John Haygarth, Smallpox and Religious Dissent in Eighteenth-Century England." In *The Medical Enlightenment of the Eighteenth Century*, ed. Andrew Cunningham and Roger French, 217–53. Cambridge: Cambridge University Press, 1990.
Lohff, Brigitte. "Zur Geschichte der Lehre von der Lebenskraft." *Clio Medica* 16 (1981): 101–12.
- *Die Suche nach der Wissenschaftlichkeit der Physiologie in der Zeit der Romantik: Ein Beitrag zur Erkenntnisphilosophie in der Medizin*. Medizin in Geschichte und Kultur 17. Stuttgart: Fischer, 1990.
Luyendijk-Elshout, Antonie. "Of Masks and Mills: The Enlightened Doctor and His Frightened Patient." In *The Languages of Psyche: Mind and Body in Enlightenment Thought*, ed. G.S. Rousseau, 186–230. Berkeley: University of California Press, 1990.
Lyon, John. "'The Science of Sciences': Replication and Reproduction in Lavater's Physiognomics." *18th-Century Studies* 40.2 (2007): 252–77.
Maehle, Andreas-Holger. *Drugs on Trial: Experimental Pharmacology and Therapeutic Innovation in the Eighteenth Century*. Amsterdam and Atlanta: Rodopi, 1999.
Mahler, Anthony. "Writing Regimens: The Dietetics of Literary Authorship in the Late German Enlightenment." Diss., University of Chicago, 2014.
Marquard, Odo. *Transzendentaler Idealismus, Romantische Naturphilosophie, Psychoanalyse*. Schriftenreihe zur philosophischen Praxis 3. Cologne: Jürgen Dinter, 1987.

McCarthy, John, et al. *The Early History of Embodied Cognition 1740–1920: The Lebenskraft-Debate and Radical Reality in German Science, Music, and Literature*. Leiden: Brill, 2016.

Menninghaus, Winfried. *Unendliche Verdopplung: Die frühromantische Grundlegung der Kunsttheorie im Begriff absoluter Selbstreflexion*. Frankfurt am Main: Suhrkamp, 1987.

Millán, Elizabeth, and John H. Smith. "Introduction – Goethe and Idealism: Points of Intersection." *Goethe Yearbook* 18 (2011): 3–9.

Miller, Elaine P. *The Vegetative Soul: From Philosophy of Nature to Subjectivity in the Feminine*. Albany: State University of New York Press, 2002.

Mischer, Sibille. *Der verschlungene Zug der Seele: Natur, Organismus und Entwicklung bei Schelling, Steffens und Oken*. Würzburg: Königshausen & Neumann, 1997.

Mitchell, Robert. *Experimental Life: Vitalism in Romantic Science and Literature*. Baltimore: Johns Hopkins University Press, 2013.

Mocek, Reinhard. *Johann Christian Reil (1759–1813): Das Problem des Übergangs von der Spätaufklärung zur Romantik in Biologie und Medizin in Deutschland*. Frankfurt am Main: Peter Lang, 1995.

Moiso, Francesco. "Magnetismus, Elektrizität, Galvanismus." In Friedrich Wilhelm Schelling, *Historisch-Kritische Ausgabe*, Reihe 1, Ergänzungsband zu Werke Band 5 bis 9: Wissenschaftshistorischer Bericht zu Schellings Naturphilosophischen Schriften 1797–1800, 165–372. Stuttgart: Frommann-Holzboog, 1994.

Möller, Hans-Jürgen. *Die Begriffe "Reizbarkeit" und "Reiz": Konstanz und Wandel ihres Bedeutungsgehaltes sowie die Problematik ihrer exakten Definition*. Stuttgart: Fischer, 1975.

Morris, David B. *The Culture of Pain*. Berkeley: University of California Press, 1991.

Morton, Timothy. *Hyperobjects: Philosophy and Ecology after the End of the World*. Minneapolis: University of Minnesota Press, 2013.

Morus, Iwan Rhys. *When Physics Became King*. Chicago: University of Chicago Press, 2005.

Müller, Lothar. "Die 'Feuerwissenschaft': Romantische Naturwissenschaft und Anthropologie bei Johann Wilhelm Ritter." In *Der ganze Mensch: Anthropologie und Literatur im 18. Jahrhundert*, ed. Hans-Jürgen Schings et al., 260–83. Stuttgart: Metzler, 1994.

Nakai, Ayako. "Poesie und Poetik bei Novalis und die Signaturenlehre der Naturmystik." In *Novalis: Poesie und Poetik*, ed. Herbert Uerlings, 185–99. Tübingen: Niemeyer, 2004.

Nassar, Dalia. "'Idealism is nothing but genuine empiricism': Novalis, Goethe, and the Ideal of Romantic Science." *Goethe Yearbook* 18 (2011): 67–95.
– *The Romantic Absolute: Being and Knowledge in Early German Romantic Philosophy, 1795–1804.* Chicago: University of Chicago Press, 2014.
Neubauer, John. *Bifocal Vision: Novalis' Philosophy of Nature and Disease.* Chapel Hill: University Press of North Carolina, 1971.
– "Intellektuelle, intellektuale und asthetische Anschauung: Zur Entstehung der romantischen Kunstauffassung." *Deutsche Vierteljahrsschrift für Literaturwissenschaft und Geistesgeschichte* 46 (1972): 294–319.
– *Symbolismus und symbolische Logik: Die Idee der ars combinatoria in der Entwicklung der modernen Dichtung.* Munich: Fink, 1978.
– "Zwischen Natur und mathematischer Abstraktion: Der Potenzbegriff in der Frühromantik." In *Romantik in Deutschland: Ein interdisziplinäres Symposium*, ed. Richard Brinkmann, 175–86. Stuttgart: Metzler, 1978.
Neuburger, Max. *Die Lehre von der Heilkraft der Natur im Wandel der Zeiten.* Stuttgart: Ferdinand Enke, 1926.
Neumann, Gerhard. *Ideenparadiese: Untersuchungen zur Aphoristik von Lichtenberg, Novalis, Friedrich Schlegel und Goethe.* Munich: Fink, 1976.
Neumann, Josef N. "Christoph Wilhelm Hufeland (1762–1836)." Vol. 1 of *Klassiker der Medizin*, ed. Dietrich von Engelhart et al., 339–59. 2nd ed., 2 vols. Munich: Beck, 1995.
Noll, Alfred. *Die "Lebenskraft" in den Schriften der Vitalisten und ihrer Gegner.* Leipzig: Voigtländer, 1914.
Nussbaum, Felicity. *The Autobiographical Subject: Gender and Ideology in 18th-Century England.* Baltimore: Johns Hopkins University Press, 1989.
Packham, Catherine. *Eighteenth-century Vitalism: Bodies, Culture, Politics.* New York: Palgrave Macmillan, 2012.
Paul, Norbert, and Thomas Schlich, eds. *Medizingeschichte: Aufgaben, Probleme, Perspektiven.* Frankfurt: Campus, 1998.
Pfeiffer, Klaus. *Medizin der Goethezeit: Christoph Wilhelm Hufeland und die Heilkunst des 18. Jahrhunderts.* Cologne: Böhlau, 2000.
Pias, Claus, ed. *Abwehr. Modelle – Strategien – Medien.* Bielefeld: Transcript, 2009.
Pickstone, John V. *Ways of Knowing: A New History of Science, Technology and Medicine.* Chicago: University of Chicago Press, 2001.
Pilloud, Séverine. "Interpretationsspielräume und narrative Autorität im biographischen Krankheitsbericht." In *Krankheit in Briefen im deutschen und französischen Sprachraum. 17.–21. Jahrhundert*, ed. Martin Dinges and Vincent Barras, 45–65. *Medizin, Gesellschaft und Geschichte: Jahrbuch des Instituts für*

Geschichte der Medizin der Robert Bosch Stiftung. Suppl. 29. Stuttgart: Franz Steiner, 2007.

Pollack-Milgate, Howard M. "'Gott ist bald 1 · ∞ – bald 1/∞ – bald 0': The Mathematical Infinite and the Absolute in Novalis." *Seminar* 51.1 (2015): 50–70.

Porter, Roy. "Barely Touching: A Social Perspective on Mind and Body." In *The Languages of Psyche: Mind and Body in Enlightenment Thought*, ed. G.S. Rousseau, 45–80. Berkeley: University of California Press, 1990.

– *The Greatest Benefit to Mankind: A Medical History of Humanity*. New York: W.W. Norton, 1997.

– "Medical Science." In *The Cambridge History of Medicine*, ed. Roy Porter, 136–75. Cambridge: Cambridge University Press, 2006.

– "What Is Disease?" In *The Cambridge History of Medicine*, 71–102.

Potter, Edward. "'Kranke Frauen': Hypochondriac Women in Comedies by C. F. Gellert and L.A.V. Gottsched." *Orbis Litterarum* 70.4 (2015): 263–305.

Pradeu, Thomas. *The Limits of the Self: Immunology and Biological Identity*. Oxford: Oxford University Press, 2012.

Rajan, Tilottama. "Excitability: The (Dis)Organization of Knowledge from Schelling's *First Outline* (1799) to *Ages of the World* (1815)." *European Romantic Review* 21.3 (2010): 309–25.

Reill, Peter Hanns. *Visualizing Nature in the Enlightenment*. Berkeley: University of California Press, 2005.

Rey, Roselyne. *The History of Pain*. Cambridge, MA: Harvard University Press, 1995.

– "Psyche, Soma, and the Vitalist Philosophy of Medicine." In *Psyche and Soma: Physicians and Metaphysicians on the Mind-Body Problem from Antiquity to Enlightenment*, ed. John P. Wright and Paul Potter, 255–65. Oxford: Claredon, 2000.

Rheinberger, Hans-Jörg. *Experiment. Differenz. Schrift: Zur Geschichte epistemischer Dinge*. Marburg: Basiliken-Presse, 1992.

– *Historische Epistemologie: Zur Einführung*. Hamburg: Junius, 2007.

Richards, Robert. *The Romantic Conception of Life: Science and Philosophy in the Age of Goethe*. Chicago: University of Chicago Press, 2002.

Richter, Klaus. "Zur Methodik der naturwissenschaftlichen Forschens bei Johann Wilhelm Ritter." In *"Fessellos durch die Systeme": Frühromantisches Naturdenken im Umfeld von Arnim, Ritter und Schelling*, ed. Walther Ch. Zimmerli, Klaus Stein, and Michael Gerten, 317–29. Stuttgart–Bad Cannstatt: Frommann-Holzboog, 1997.

Rieger, Stefan. "Die Kybernetik des Menschen: Steuerungswissen um 1800." In *Poetologien des Wissens um 1800*, ed. Joseph Vogl, 97–119. Munich: Fink, 1999.

Risse, Guenther B. "The History of John Brown's Medical System in Germany during the Years 1790–1806." Diss., University of Chicago, 1971.
- "Kant, Schelling, and the Early Search for a Philosophical 'Science' of Medicine in Germany." *Journal of the History of Medicine* 27.2 (1972): 145–58.
- "'Philosophical' Medicine in Nineteenth-Century Germany: An Episode in the Relations between Philosophy and Medicine." *Journal of Medicine and Philosophy* 1.1 (1976): 72–92.

Rommel, Gabriele. "Romanticism and Natural Science." In *The Literature of German Romanticism*, ed. Dennis F. Mahoney, 209–27. The Camden House History of German Literature 8. Rochester, New York: Camden House, 2004.

Rommel, Gabriele, ed. *Novalis: Geheimnisvolle Zeichen: Alchemie, Magie, Mystik und Natur bei Novalis*. Leipzig: Edition Leipzig, 1998.

Rothschuh, Karl. E. *Konzepte der Medizin in Vergangenheit und Gegenwart*. Stuttgart: Hippokrates, 1978.
- "Naturphilosophische Konzepte der Medizin aus der Zeit der deutschen Romantik." In *Romantik in Deutschland: Ein interdisziplinäres Symposion*, ed. Richard Brinkmann, 243–66. Stuttgart: Metzler, 1978.
- *Physiologie: Der Wandel ihrer Konzepte, Probleme und Methoden vom 16. bis 19. Jahrhundert*. Freiburg, Munich: Karl Alber, 1968.

Rousseau, G.S., and Roy Porter. "Introduction: Toward a Natural History of Mind and Body." In *The Languages of Psyche: Mind and Body in Enlightenment Thought*, ed. G.S. Rousseau, 3–44. Berkeley: University of California Press, 1990.

Rudolph, Gerhard. "Leitgedanken der Diagnostik und Semeiotik in der französischen Medizin des 18. und frühen 19. Jahrhunderts." In *Medizinische Diagnostik in der Geschichte und Gegenwart: Festschrift für Heinz Goerke*, ed. Christa Habrich et al., 270–81. Munich: Fritsch, 1978.

Rusnock, Andrea. "Making Sense of Vaccination c. 1800." In *Crafting Immunity: Working Histories of Clinical Immunology*, ed. Kenton Kroker et al., 17–27. Aldershot: Ashgate, 2008.

Ruston, Sharon. *Shelley and Vitality*. New York: Palgrave Macmillan, 2005.

Sarasin, Philipp. *Reizbare Maschinen: Eine Geschichte des Körpers 1765–1914*. Frankfurt: Suhrkamp, 2001.

Sarasin, Philipp, et al. *Bakteriologie und Moderne: Studien zur Biopolitik des Unsichtbaren 1870–1920*. Frankfurt: Suhrkamp. 2007.

Scheer, Monique. "Topografien des Gefühls." In *Gefühlswissen: Eine lexikalische Spurensuche in der Moderne*, ed. Ute Frevert et al., 41–64. Frankfurt: Campus, 2011.

Schiffter, Roland. ". . . *ich habe immer klüger gehandelt . . . als die philisterhaften Ärzte . . .*": *Romantische Medizin im Alltag der Bettina von Arnim – und anderswo*. Würzburg: Königshausen & Neumann, 2006.

Schipperges, Heinrich. "Grundzüge einer 'polarischen' Medizin bei Novalis." *Antaios* 7 (1965): 196–208.

– "Krankwerden und Gesundsein bei Novalis." In *Romantik in Deutschland: Ein interdisziplinäres Symposion*, ed. Richard Brinkmann, 226–42. Stuttgart: Metzler, 1978.

Schlächter, H., and K.-E. Bühler. "Theorien über Traum und Träumen im Deutschland des 19. Jahrhunderts. *Fundamenta Psychiatrica* 17 (2003): 35–42.

Schnyder, Peter. *Die Magie der Rhetorik: Poesie, Philosophie und Politik in Friedrich Schlegels Frühwerk*. Paderborn, Munich, Vienna, Zurich: Schöningh, 1999.

Schott, Heinz, ed. *Meilensteine der Medizin*. Dortmund: Harenberg, 1996.

Schrader, Hans-Jürgen, and Katharine Weder, eds. *Von der Pansophie zur Weltweisheit: Goethes analogisch-philosophische Konzepte*. Tübingen: Niemeyer, 2004.

Schwanitz, Hans. *Die Theorie der praktischen Medizin zu Beginn des 19. Jahrhunderts: Eine historische und wissenschaftstheoretische Untersuchung anhand des "Journals der praktischen Arzneykunde und Wundarzneykunde" von Ch. W. Hufeland*. Cologne: Pahl-Rugenstein, 1979.

Seamon, David, and Arthur Zajonc, eds. *Goethe's Way of Science: A Phenomenology of Nature*. Albany: State University of New York Press, 1998.

Seigel, Jerrold. *The Idea of the Self: Thought and Experience in Western Europe since the Seventeenth Century*. Cambridge: Cambridge University Press, 2005.

Shapiro, Johanna, et al. "Medical Humanities and Their Discontents: Definitions, Critiques, and Implications." *Academic Medicine* 82.2 (2009): 192–98.

Smith, John H. "Friedrich Schlegel's Romantic Calculus: Reflections on the Mathematical Infinite around 1800." In *The Relevance of Romanticism: Essays on German Romantic Philosophy*, ed. Dalia Nassar, 239–57. Oxford: Oxford University Press, 2014.

– "Leibniz Reception around 1800: Monadic Vitalism and Aesthetic Harmony." In *Religion, Reason, and Culture in the Age of Goethe*, ed. Elisabeth Krimmer and Patricia Anne Simpson, 209–43. Rochester, New York: Camden House, 2013.

Smith, Sidonie, and Julia Watson. *Reading Autobiography: A Guide for Interpreting Life Narratives*. Minneapolis: University of Minnesota Press, 2001.

Sohni, Hans. *Die Medizin der Frühromantik: Novalis' Bedeutung für den Versuch einer Umwertung der 'Romantischen Medizin.'* Freiburger Forschungen zur Medizingeschichte 2 (new series). Freiburg: Schultz, 1973.

Solhdju, Katrin. *Selbstexperimente: Die Suche nach der Innenperspektive und ihre epistemologischen Folgen*. Munich: Fink, 2011.

Speler, Ralf-Torsten, ed. *Das geheimnisvolle Organ: Die Vorstellung über Hirn und Seele von Johann Christian Reil bis heute*. Halle: Martin-Luther-Universität Halle-Wittenberg, 2013.

Stadler, Ulrich. "'Ich lehre nicht, ich erzähle': Über den Analogiegebrauch im Umkreis der Romantik." *Athenäum* 3 (1993): 83–105.

Stafford, Barbara Maria. *Visual Analogy: Consciousness as the Art of Connecting*. Cambridge, MA: MIT Press, 1999.

Stalfort, Jutta. *Die Erfindung der Gefühle: Eine Studie über den historischen Wandel menschlicher Emotionalität (1750–1850)*. Bielefeld: Transcript, 2013.

Steger, Florian, ed. *Johann Christian Reil: Universalmediziner, Stadtphysikus, Wegbereiter von Psychiatrie und Neurologie*. Gießen: Psychosozial-Verlag, 2014.

Steigerwald, Joan. "Epistemologies of Rupture: The Problem of Nature in Schelling's Philosophy." *Studies in Romanticism* 41 (Winter 2002): 554–84.

– "Figuring Nature: Ritter's Galvanic Inscriptions." *European Romantic Review* 18.2 (2007): 255–63.

– "Instruments of Judgment: Inscribing Organic Processes in Late 18th-Century Germany." *Studies in History and Philosophy of Biological and Biomedical Sciences* 33 (2002): 79–131.

– "Kant's Concept of Natural Purpose and the Reflecting Power of Judgment." *Studies in History and Philosophy of Biological and Biomedical Sciences* 37 (2006): 712–34.

– "Rethinking Organic Vitality in Germany at the Turn of the Nineteenth Century." In *Vitalism and the Scientific Image in Post-Enlightenment Life Science, 1800–2010*, ed. Sebastian Normandin and Charles T. Wolfe, 51–75. History, Philosophy and Theory of the Life Sciences 2. Dordrecht: Springer, 2013.

– "Treviranus' Biology: Generation, Degeneration and the Boundaries of Life." In *Reproduction, Race, and Gender in Philosophy and the Early Life Sciences*, ed. Suzanne Lettow, 105–27. Albany: SUNY Press, 2015.

Steigerwald, Jörn, and Daniela Watzke, eds. *Reiz Imagination Aufmerksamkeit: Erregung und Steuerung von Einbildungskraft im klassischen Zeitalter (1680–1830)*. Würzburg: Königshausen & Neumann, 2003.

Steinke, Hubert. *Irritating Experiments: Haller's Concept and the European Controvesy on Irritability and Sensibility, 1740–90*. Amsterdam: Rodopi, 2005.

Stelzig, Eugene. *The Romantic Subject in Autobiography: Rousseau and Goethe*. Charlottesville: University Press of Virginia, 2000.

Stolberg, Michael. *Experiencing Illness and the Sick Body in Early Modern Europe*. Basingstoke, Hampshire: Palgrave Macmillan, 2011.

- "'Mein äskulapisches Orakel!' Patientenbriefe als Quelle einer Kulturgeschichte der Körper- und Krankheitserfahrung im 18. Jahrhundert." *Österreichische Zeitschrift für Geschichtswissenschaften* 7 (1996): 385–404.
- "Patientenbriefe in vormoderner Medikalkultur." In *Krankheit in Briefen im deutschen und französischen Sprachraum. 17.–21. Jahrhundert*, ed. Martin Dinges and Vincent Barras, 23–33. *Medizin, Gesellschaft und Geschichte: Jahrbuch des Instituts für Geschichte der Medizin der Robert Bosch Stiftung*. Suppl. 29. Stuttgart: Franz Steiner, 2007.

Strickland, Stuart Walter. "The Ideology of Self-Knowledge and the Practice of Self-Experimentation." *18th-Century Studies* 31.4 (1998): 453–71.

Szondi, Peter. *Poetik und Geschichtsphilosophie II*. Frankfurt: Suhrkamp, 1974.

Tanner, Jakob. "Körpererfahrung, Schmerz und die Konstruktion des Kulturellen." *Historische Anthropologie* 2 (1994): 488–502.

Tantillo, Astrida Orle. *The Will to Create: Goethe's Philosophy of Nature*. Pittsburgh: University of Pittsburgh Press, 2002.

Tatar, Maria. *Spellbound: Studies on Mesmerism and Literature*. Princeton: Princeton University Press, 1978.

Tauber, Alfred. *The Immune Self: Theory or Metaphor?* Cambridge: Cambridge University Press, 1994.

Taylor, Charles. *Hegel*. Cambridge: Cambridge University Press, 1975.
- *Sources of the Self: The Making of the Modern Identity*. Cambridge, MA: Harvard University Press, 1989.

Thums, Barbara. "Aufmerksamkeit: Zur Ästhetisierung eines anthropologischen Paradigmas im 18. Jahrhundert." In *Reiz Imagination Aufmerksamkeit: Erregung und Steuerung von Einbildungskraft im klassischen Zeitalter (1680–1830)*, ed. Jörn Steigerwald and Daniela Watzke, 55–74. Würzburg: Königshausen & Neumann, 2003.

Tobin, Robert. *Doctor's Orders: Goethe and Enlightenment Thought*. Lewisburg, PA: Bucknell University Press, 2001.

Tsouyopoulos, Nelly. *Asklepios und die Philosophen: Paradigmawechsel in der Medizin im 19. Jahrhundert*. Ed. Claudia Wiesemann, Barbara Bröker, und Sabine Rogge. Stuttgart–Bad Cannstatt: Frommann-Holzboog, 2008.
- "The Influence of John Brown's Ideas in Germany." In *Brunonianism in Britain and Europe*, ed. W.F. Bynum and Roy Porter, 63–74. London: Wellcome Institute for the History of Medicine, 1988.

Türk, Johannes. "Freuds Immunologien des Psychischen." *Poetica* 38.1–2 (2006): 167–88.
- "Homo immunis. Zur Genese und Topologie des modernen Menschen in der Immunologie." In *Engineering Life: Narrationen vom Menschen in Biomedizin,*

Kultur und Literatur, ed. Claudia Breger, Irmele Krüger-Fürhoff, and Tanja Nusser, 71–88. Berlin: Kulturverlag Kadmos, 2008.

Uerlings, Herbert. *Friedrich von Hardenberg, genannt Novalis: Werk und Forschung*. Stuttgart: Metzler, 1991.

Van den Berg, Hein. "Kant's Conception of Proper Science." *Synthese* 183 (2011): 7–26.

Vigarello, Georges. *Wasser und Seife, Puder und Parfüm: Geschichte der Körperhygiene seit dem Mittelalter*. Nachwort von Wolfgang Kaschuba. Frankfurt am Main: Campus Verlag, 1988.

Vila, Anne C. *Enlightenment and Pathology: Sensibility in the Literature and Medicine of Eighteenth-Century France*. Baltimore: Johns Hopkins University Press, 1998.

Vlasopoulos, Michail. "Spinoza's God in Goethe's Leaf: The Spinozistic Foundations of Goethean Morphology." Paper presented at the 7th Graduate Conference of the Centre for Research on Religion (CREOR). McGill University, Montreal, September 2015.

Wallen, Martin. *City of Health, Fields of Disease: Revolution in the Poetry, Medicine, and Philosophy of Romanticism*. Aldershot: Ashgate, 2004.

Wasmuth, Ewald. "Novalis' Beitrag zu einer 'Physik in einem höheren Stile.'" *Neue Schweizer Rundschau* 18 (1950–1): 533–46.

Watzke, Daniela. "Hirnananatomische Grundlagen der Reizleitung und die 'bewusstlose Sensibilität' im Werk des Hallenser Klinikers Johann Christian Reil." In *Reiz Imagination Aufmerksamkeit: Erregung und Steuerung von Einbildungskraft im klassischen Zeitalter (1680–1830)*, ed. Jörn Steigerwald and Daniela Watzke, 248–67. Würzburg: Königshausen & Neumann, 2003.

Weder, Katharine. *Kleists magnetische Poesie: Experimente des Mesmerismus*. Göttingen: Wallstein, 2008.

Wellbery, David E. *Lessing's "Laocoon": Semiotics and Aesthetics in the Age of Reason*. Anglica Germanica Series 2. Cambridge: Cambridge University Press, 1984.

Wetzels, Walter. *Johann Wilhelm Ritter: Physik im Wirkungsfeld der deutschen Romantik*. Berlin: de Gruyter, 1973.

Wiesemann, Claudia. *Die heimliche Krankheit: Eine Geschichte des Suchtbegriffs*. Medizin und Philosophie 4. Stuttgart–Bad Cannstatt: Frommann-Holzboog, 2000.

Wiesing, Urban. *Kunst oder Wissenschaft? Konzeptionen der Medizin in der deutschen Romantik*. Stuttgart: Frommann-Holzboog, 1995.

Williams, Elizabeth A. *The Physical and the Moral: Anthropology, Physiology, and Philosophical Medicine in France, 1750–1850*. Cambridge: Cambridge University Press, 1994.

Winter, Alison. *Mesmerized: Power of Mind in Victorian Britain*. Chicago: University of Chicago Press, 1998.

Wood, David W. "The 'Mathematical' *Wissenschaftslehre*: On a Late Fichtean Reflection of Novalis." In *The Relevance of Romanticism: Essays on German Romantic Philosophy*, ed. Dalia Nassar, 258–72. Oxford: Oxford University Press, 2014.

Wright, John P. "Substance versus Function Dualism in Eighteenth-Century Medicine." In *Psyche and Soma: Physicians and Metaphysicians on the Mind-Body Problem from Antiquity to Enlightenment*, ed. John P. Wright and Paul Potter, 267–54. Oxford: Clarendon, 2000.

Zelle, Carsten. "Erfahrung, Ästhetik und mittleres Maß: Die Stellung von Unzer, Krüger und E.A. Nicolai in der anthropologischen Wende um 1750." In *Reiz Imagination Aufmerksamkeit: Erregung und Steuerung von Einbildungskraft im klassischen Zeitalter (1680–1830)*, ed. Jörn Steigerwald and Daniela Watzke, 204–24. Würzburg: Königshausen & Neumann, 2003.

Zelle, Carsten, ed. *"Vernünftige Ärzte": Hallesche Psychomediziner und die Anfänge der Anthropologie in der deutschprachigen Aufklärung*. Tübingen: Niemeyer, 2001.

Ziolkowski, Theodore. *Das Wunderjahr in Jena: Geist und Gesellschaft 1794/95*. Stuttgart: Klett-Cotta, 1998.

Zumbusch, Cornelia. *Die Immunität der Klassik*. Frankfurt: Suhrkamp, 2011.

Index

addiction, 31, 176n9
Agamben, Giorgio, 11; on paradigm, 10
alchemy, 48, 130, 165n29
Aldini, Giovanni, 86
allopathy, 18, 51, 67, 112, 137; chronic disease caused by, 169n25; definition of, 16; use of opium in, 32
Amrine, Frederick, 180n39, 183n7, 183n11; on Spinoza, 152, 183n5, 183n13
anamnesis, 7, 16, 71, 74, 77; attention to *Gemüt* in the, 87, 138; taking precedence over diagnosis, 100
anima, theory of, 42
animal magnetism, 9, 116, 130–3. *See also* mesmerism
Arnim, Achim von, 97
Arnim, Bettina von, 4, 138, 143, 159n3
Arz, Maike, 144
Authenrieth, Johann Ferdinand, 116

Baader, Franz von, 57
Bach, Friedrich Christian, 81, 169n22, 171n47
Barkhoff, Jürgen, 182n51
Barthez, Paul-Joseph, 118–19
Baschin, Marion, 75, 77, 170n33, 171n41
Baudelaire, Charles, 56
Bayr, Georg, 162n4, 169n19
Bell, Matthew, 173n59
Benjamin, Walter, 99; "Lehre vom Ähnlichen" ("Doctrine of the Similar"), 48–58 passim, 165n31
Bennett, Jane, 178n20
Bergengruen, Maximilian, 165n28
Berrington, Joseph, 15
Bertalanffy, Ludwig von, 139
Bichat, Xavier, 42, 117
Bildung, concept of, 60–1, 66
Bildungsroman, 61, 82
Bildungstrieb, 116
Bleul, Gerhard, 52
Blumenbach, Friedrich, 116–18, 129
Boerhaave, Herman, 41–2, 81, 91, 118, 164n20
Böhme, Jacob, 166n41
Bönninghausen, Clemens von, 4, 74–7 passim, 143, 171n41; and Annette von Droste-Hülshoff, 4, 79, 101, 171n42

Bordeu, Theophile de, 42
Borgards, Roland, 86–7, 173n60
Botsch, Walter, 177n16, 179n32
Boyle, Robert, 4l1
Brandis, Johann Dietrich, 86
Breithaupt, Fritz, 172n48
Brockmeyer, Bettina, 101, 170n34, 172n52, 175n77, 180n40
Broman, Thomas Hoyt, 161n16, 162n7, 164n20, 169n29, 178n25; on fevers, 169n19; on Hufeland, 165n27
Broussais, François-Joseph-Victor, 168n16
Brown, John, 44, 47, 64–6, 168n16, 169n20, 176n9; comparison with Hahnemann, 22, 84, 108, 121–3, 137, 168n17, 173n64; *Elementa medicinae*, 64; on *Erregbarkeit*, 146; Hufeland on, 31, 120–1; influence on Schelling, 168n14; legacy of, 178n25; Novalis on, 65, 109
Burdach, Karl Friedrich, 86–7, 176n7; on Hufeland, 109
Busche, Jens, 74–5, 170n35
Bütschli, Otto, 120

Cabanis, Pierre Georges, 42
calculus, 126–7, 140, 160n6, 183n6
Carus, Carl Gustav, 120
case study, 69, 78; case history, 69, 85, 90, 169n24, 171n42
chirurgy, 86
Classicism, 20, 138
coffee, 7, 75, 77, 89, 136, 169n19. *See also* Hahnemann, "Der Kaffee"
Coleridge, Samuel Taylor, 142, 150; "Theory of Life," 148
Comte, Auguste, 93
contraria contrariis, 27–8, 33, 35
Coulter, Harris, 23, 52, 178n23

Crumpe, Samuel, 31
Cullen, William, 31, 37, 90, 162n23, 163n14; *Materia medica*, 15; on neurosis, 81

Daiber, Jurgen, 92, 96–7; on Novalis, 174n72
Daston, Lorraine, 93–4
Davy, Humphry, 175n80
Deleuze, Gilles, "Body without Organs," 99; *Expressionism in Philosophy*, 148; *Spinoza*, 147, 150; *A Thousand Plateaus*, 50, 166n32; *What Is Philosophy*, 183n8
Descartes, 81–2, 95, 118, 148
d'Hervilly, Mélanie, 19
Diderot, Denis, 82
dietetics, 9, 43, 86, 99, 107, 130, 182n50; Hahnemann's writings on, 170n35; purpose of 75; and self-surveillance, 74, 170n32, 170n34
dilution, 13, 48, 103–50 passim, 176n8, 177n13
Dinges, Martin, 72, 74, 162nn25–6, 169n27, 170n31, 170n34, 175n79
dosaging, 68, 70, 169n20; Hahnemann's instructions for, 113, 177n13
Droste-Hülshoff, Annette von, 4, 79–80, 101, 171n42
Duden, Barbara, 25, 71, 99, 169n28; on *Alltagsgeschichte*, 24
Dugeon, Robert Ellis, 22
Dyck, Martin, 141
dynamization, 104, 111, 140. *See also Potenzierung*

Eichendorff, Joseph von, 55–6
Elgin, Lord Thomas Bruce, 19, 101
Empfänglichkeit, 138

empiricism, 12, 19–25 passim, 32, 42, 91–7 passim, 119, 146; Boerhaave on, 91; Deleuze on radical, 183n8; Goethe on, 95, 97; Hahnemann's use of the term, 173n64; Heinroth on Hahnemann's, 46; in Kant, 129, 180n40; in Novalis, 173n65, 174n72; in Ritter, 97, 175n73; and Romantic science, 147. See also *Erfahrung*

Engelhardt, Dietrich von, 125, 147–8, 176n11, 182n51

Enlightenment, 45, 82; Enlightenment rationalism 20, 54; Hahnemann as a thinker of, 14, 19, 58, 152, 160n7, 180n40

Erfahrung (experience), 96, 120, 154–6, 168n14, 173n64; *Erfahrungsseelenkunde* (experiential psychology), 81; *Erfahrungsvitalitätskunde* (empirical knowledge of vitality), 118; *Erfahrungswissenschaft* (science of experience), 91

Eschenmayer, Carl August von, 132

Fichte, Johann Gottlieb, 12, 91; "Ueber die Bestimmung des Gelehrten" ("On the Vocation of the Scholar"), 138; on the self, 145

Fischbach-Sabel, Ute, 23, 39, 41, 74, 77, 87

Förster, Eckart, 180n39

Foucault, Michel, 18, 99; *Birth of the Clinic*, 62; *Order of Things*, 45, 47

Frank, Manfred, on *Selbstgefühl*, 82

Frazer, James, *The Golden Bough*, 132

French, Roger, 42

Galen, 30, 62, 64, 72, 167n4; Galenist therapies, 44

Galison, Peter, 93–4
Gall, Franz Joseph, 81
Galvani, Luigi, 116–17, 125
galvanism, 130
Gantenbein, Urs Leo, 165n29
Gemüt (*Gemüth*), 53, 79–86 passim, 138, 141, 172n51, 172nn53–4; *Gemütsbildung*, 66
genius, 147, 173n64, 175n79, 178n26; cult of, 102, 157; Hahnemann as, 8, 24, 102; nature as, 121; Novalis on, 97

Genneper, Thomas, 170n36
Geyer-Kordesch, Johanna, 178n21
Gigante, Denise, 117
Gmelin, Eberhard, 130
Goethe, Johann Wolfgang von, 10, 20, 60, 81, 91, 118, 142, 146–56 passim, 183n6; on analogy, 56, 166n39; *anschauende Urteilskraft* (intuitive power of judgment), 129, 152–3, 183n11; correspondence, 94–6, 148, 154, 183n7; "Erfahrungen der höheren Art," 156; on experimentation, 97, 167n3, 174n71; *Faust*, 8–9, 107, 144; on a Hahnemannian diet, 7; on homeopathy, 3–4, 7, 109, 174n70; *Lila*, 35–6; on matter and spirit, 125; *Maxims and Reflections*, 96; *Die Metamorphose der Pflanzen* (*The Metamorphosis of Plants*), 152; on morphology, 95, 117, 152, 177n18; and pharmakon, 109; reception of Kant, 153, 156, 180n39; and Spinoza, 147–52, 183n5; on a stoic response to illness, 85, 138; *Urfaust*, 152; *Urpflanze* (archetypal plant), 95–6, 153–4; "Der Versuch als Vermittler von Objekt und Subjekt" ("The Experiment as

Mediator between Object and Subject"), 150; *Wilhelm Meisters Lehrjahre* (*Wilhelm Meister's Years of Apprenticeship*), 61, 174n70
Goldstein, Jan, 63, 170n37, 172n58
Guattari, Felix, *A Thousand Plateaus*, 50, 166n32

Hadot, Pierre, on "Orphic Tradition," 142
Haehl, Richard, 41, 104, 147, 160n10, 165n29, 177n13, 179n28
Hahnemann, Samuel, biographies on, 14–15, 160n10; correspondence with, 24, 50, 63, 72, 75, 101, 171n38; on God, 14, 47, 57, 91, 150; his life, 9, 14–19, 24; on the LM potency (Q-potency), 20, 104, 175n1; on medical practices of his day, 27, 78, 86; on mental illness, 78, 85–6; on *Naturheilgesetz*, 16, 143, 154–5; on palliative medicine, 29, 30–3, 35, 84, 87, 108; and Peruvian bark, 16, 31, 97, 162n4, 169n19, 174n67; on polychrests, 76; rules for conducting provings, 89–90; on the secondary effect, 30–4, 123, 163n8, 176n9; on self-experimentation, 13, 90–1, 125, 157, 175n80, 182n51
– collections and published works: "Aeskulap auf der Wagschaale" ("Aesculapius in the Balance"), 18, 27, 29, 64–5, 67, 165n29, 167n4; "Die Allöopathie. Ein Wort der Warnung an Kranke jeder Art" ("Allopathy: A Word of Warning to All Sick Persons"), 18–19, 68, 108, 143; "Die Ansteckungsart der asiatischen Cholera" ("The Mode of Propagation of Asiatic Cholera"), 15; "Der ärztliche Beobachter" ("The Medical Observer"), 93–4; "Auszug eines Briefs an einen Arzt von hohem Range" ("Necessity of a Regeneration of Medicine"), 9, 18, 28, 69, 135; "Beleuchtung der Quellen der gewöhnlichen Materia medica" ("The Sources of the Common Materia Medica"), 64, 69, 88, 105, 124; "Bemerkungen über das Scharlachfieber" ("Observations on the Scarletfever"), 114; "Cases Illustrative of Homeopathic Practice," excerpted from *Reine Arzneimittellehre*, 69–70; *Die chronischen Krankheiten* (*The Chronic Diseases*), 17, 23, 38, 70, 79, 83, 88, 112–13, 122; *Fragmenta de viribus medicamentorum*, 17, 37–40, 88; "Fragmentarische Bemerkungen zu Browns *Elements of Medicine*" ("Fragmentary Observations on Brown's *Elements of Medicine*"), 27, 64; *Freund der Gesundheit* (*The Friend of Health*), 15, 74, 135–6; "Gedanken bey Gelegenheit des Mittels gegen die Folgen des Bisses toller Hunde" ("On a Proposed Remedy for Hydrophobia"), 136; "Geist der neuen Heillehre" ("Spirit of the Homoeopathic Doctrine of Medicine"), 108, 124, 180n36; "Heilkunde der Erfahrung" ("The Medicine of Experience"), 17–18, 29, 33–5, 37, 43, 51, 69, 94, 104–5, 110, 113, 121, 124, 131, 137, 162n2, 166n33, 177n14, 179n27;

"Heilung und Verhütung des Scharlach-Fiebers" ("Cure and Prevention of Scarlet-fever"), 17, 112, 135–6, 163n9, 173n64; "Der Kaffee in seinen Wirkungen" ("On the Effects of Coffee"), 30, 68, 170n35; *Krankenjournale* (*Case Journals*), 23–4, 40, 50–1, 69, 72, 75–9, 87, 99–100, 111–12, 115, 130, 164n16, 164n18, 171n40, 172n53, 181n47; "Monita über die drey gangbaren Kurarten" ("Three Current Methods of Treatment"), 18, 27, 63–5, 68, 121, 165n26, 165n29, 173n64; "Nachschrift des Herrn Hofrath Dr. S. Hahnemann" ("Remarks on the Extreme Attenuation of Homoeopathic Remedies"), 105, 107; "Old and New Systems of Medicine," excerpted from *Reine Arzneimittellehre*, 14, 18, 67–8, 71, 122, 143; *Organon der Heilkunst* (*Organon of the Art of Healing*), 15, 17–18, 20, 22–3, 28–30, 32–3, 37, 39, 41, 46, 52–3, 57–60, 62, 68, 70–1, 74, 75, 77–8, 83–8, 91–2, 98, 104–6, 112–14, 120, 122–4, 126, 130–1, 135, 140, 143, 145, 149, 150, 154, 156, 159n5, 161nn14–16, 165n25, 168n15, 169n25, 170n33, 172n54, 173n64, 179n27, 179n34, 181n43, 181n48, 182n53, 183n14; reviews of, 22, 46–7, 57–8, 78, 161n17, 179n34, 183n14; *Organon der rationellen Heilkunde* (*Organon of Rational Medical Science*), 16; *Organon-Synopse*, 121, 123, 155, 161n15; protocol books, 37, 90, 98, 115, 130, 149, 161n23, 163n14;

Reine Arzneimittellehre (*Materia Medica Pura*), 16–17, 21, 22–3, 32–3, 37–8, 45, 49–51, 69–70, 88, 93, 98, 101, 112; repertories, 37–8, 75, 98, 163n15; "Sind die Hindernisse der Gewißheit und Einfachheit der practischen Arzneykunde unübersteiglich?" ("Are the Obstacles to Certainty and Simplicity in Practical Medicine Insurmountable?"), 17, 18, 74; "Two Cases from Hahnemann's Note Book," 70; "Ueber den jetzigen Mangel außereuropäischer Arzneyen" ("On the Present Want of Foreign Medicine"), 30; "Ueber den Werth der speculativen Arzneysysteme" ("On the Value of the Speculative Systems of Medicine"), 8, 18, 51, 70, 165n26, 180n36; "Ueber die Kraft kleiner Gaben der Arzneien überhaupt und der Belladonna insbesondre" ("On the Power of Small Doses of Medicine"), 136–7; "Ueber die Lieblosigkeit gegen Selbstmörder" ("On the Uncharitableness Towards Suicides"), 132; "Ueber die Surrogate ausländischer Arzneyen" ("On Substitutes for Foreign Drugs"), 69; "Versuch über ein neues Prinzip zur Auffindung der Heilkräfte der Arzneisubstanzen" ("Essay on a New Principle for Ascertaining the Curative Powers of Drugs"), 12, 18, 27–30, 103, 108, 111; "Vorrede" ("A Preface"), 17, 121, 175n2; "Vorstellung an Eine hohe

Behörde" ("Representation to a Person High in Authority"), 51; "Wie können kleine Gaben so sehr verdünnter Arznei noch große Kraft haben?" ("How Can Small Doses Still Possess Great Power?"), 105
Haller, Albrecht von, 43–4, 81, 120; and the Montpellier school, 177n15, 178n22; division into irritability and sensibility, 82, 108, 116, 118, 137, 176n10; *Materia medica*, 15
Hardenberg, Friedrich von. *See* Novalis
Hartmann, Franz, 89–90
Hecker, August Friedrich, 46–7
Hegel, Georg Wilhelm Friedrich, 55, 96, 124, 151; *The Science of Logic*, 127
Heinroth, Johann Christian August, 46, 86, 172n56; *Anti-Organon*, 47, 78–9
Heinz, Inge Christine, 76–7, 171n38
Heinz, Jutta, 79, 82, 166n40
Helmont, Jan Baptist van, 118
Herder, Johann Gottfried, 10, 142, 147, 154; on analogical invention, 56, 166n40; "Gott, einige Gespräche" ("God, Some Conversations"), 115–17; on *Kraft*, 82, 115, 143; on Spinoza, 142, 147; "Vom Erkennen und Empfinden der menschlichen Seele" ("On Cognition and Sensation of the Human Soul"), 66–7
Hering, Constantine, 22, 159n5
Hess, Volker, 53, 163n13, 164n24, 167n8
Hickmann, Reinhard, 77, 85

Hippocrates, 22, 61, 117–19, 122
Hoffmann, E.T.A., "Der goldne Topf" ("The Golden Pot"), 133; "Der Magnetiseur" ("The Mesmerist"), 134
Hoffmann, Friedrich, 41–2, 63
Hohenheim, Theophrastus von. *See* Paracelsus
Hölderlin, Friedrich, 138, 147
Holmes, Richard, 126, 160n8
homeopathy, etymology of, 16; definition of, 3
Hörsten, Iris von, 76–8, 111, 164n16, 164n18, 171n40
Hufeland, Christoph, 10, 82, 108, 129, 164n22, 165n27, 173n63; and John Brown, 65, 121, 137; as editor of *Journal der practischen Arzneykunde und Wundarzneykunst* (*Journal of Practical Medicine and Chirurgy*), 3, 44, 112, 118; on *Erregbarkeit*, 176n10; Hahnemann's praise and criticism of, 64, 178n24; review of Hahnemann, 38, 47, 124; *Die Kunst, das menschliche Leben zu verlängern* (*Art of Prolonging Life*), 118–19, 134–5; on *Lebenskraft*, 82, 117–24, 128, 159n1; on macrobiotics, 61, 75, 109, 118, 159n1; on mesmerism, 181n42; on natural healing, 43, 178n26; on opium addiction, 31, 176n5; on Paracelsus, 165n29
Humboldt, Alexander von, 20, 31, 60, 94, 116–17; *Kosmos*, 149; on *Naturphilosophie*, 147; and self-experimentation, 91, 97
Hume, David, 11
hypnotism, 130
hypochondria, 71, 83, 85, 111, 172n50

Index

iatrochemical (iatromechanical), 44, 86
immunology, 134, 139
infinitesimal dose, 13, 48, 106–8, 122, 128, 150
inoculation, 134–5, 137
irritability (*Reizbarkeit*), 30, 76, 131, 134, 170n32, 176n10. *See also* Haller

Jay, Martin, 173n65
Jena Romantics, the, 20, 100, 140
Jenner, Edward, 134, 181n44
Jewson, Nicholas, 167n1
Jütte, Robert, 24, 74, 160n10, 175n1

Kant, Immanuel, 11–12, 91–3, 145, 151–6 passim, 173nn64–5; on epistemological limits, 115; Goethe and, 180n39; Hahnemann and, 20, 85, 127–9, 156, 180n36, 180n40, 183n14; *Third Critique*, 128, 138
Kelley, Theresa, 117, 129, 177n18
Kent, James Tyler, 79
Kielmeyer, Carl Friedrich von, 116
Kilian, Conrad Joseph, 117
Kleist, Heinrich von, 96
Klunker, Will, 23
Krell, David Farrell, 111
Kuhn, Bernhard, 95

Lamarck, Jean-Baptiste, 124
La Mettrie, Julien Offray de, 81, 118
Lavater, Johann Christian, 46, 81, 171n46
law of minimum, 10, 13, 66, 103–31 passim, 150; definition of, 4, 103–4. *See also* dilution; dynamization; *Potenzierung*
law of similars, 10, 12, 27–58 passim; and mesmerism, 131, 163n11; definition of, 3, 27. *See also similia similibus curentur*
law of the single remedy, the, 10, 12, 60, 76; definition of, 59
laws of nature, 11, 16, 108, 133, 145
Lebenskraft (vital life force), 20, 106, 113–37 passim, 156, 160n6, 177n16, 179nn28–30; current scholarship on, 116–17, 144, 177–8nn16–20; as coined by Medicus, 115. *See also* Hufeland, on *Lebenskraft*
Leibniz, Gottfried Wilhelm, 126, 147
Lenoir, Timothy, 117
Levin, Karl-Heinz, 175n76
Linnaeus, Carl, 37
Lucae, Christian, 23, 32, 38, 90, 163n14
Luther, Martin, 9

macrobiotics, 3; definition of, 159n1. *See also under* Hufeland
Maehle, Andreas Holger, 31–2, 63, 90–1, 162n6, 167nn9–10; beginnings of pharmacology, 169n23; on new pharmacopeia, 168n12
Marc, Carl Christian Heinrich, 44, 64, 76, 110
Marcus, Adalbert Friedrich, 64
Marquard, Odo, 143
Mayr, Stefan, 111, 182n52
medical humanities, 6, 7
Medicus, Friedrich Casimir, 115, 118
Mesmer, Frank Anton, 130. *See also* animal magnetism; mesmerism
mesmerism, 81, 107, 115–16, 120; comparison to homeopathy, 129–33, 156, 160n6; use in homeopathy, 14; Hufeland on, 181n41

Millán, Elizabeth, 153
Miller, Elaine, 177n18
Mischer, Sibille, 129, 160n8
Mitchell, Robert, 117, 124
Mocek, Reinhard, 178n25
Montpellier School, 42, 119, 177n15
Moritz, Karl Philip, 81
Morton, Timothy, 178n20
Mortsch, Markus, 76–7
Morus, Iwan Rhys, 102, 126
Müller, Johannes, 86
mysticism, 54, 118

Nassar, Dalia, 183n6
natural powers of healing, 137; self-healing, 65, 121, 123
natura naturans, 20, 142, 148, 183n7
Naturphilosophie, 96, 129, 143, 146–7, 179n32, 180n40; Hufeland and Hahnemann's criticism of, 65
Neubauer, John, 110; on *Potenzierung*, 140
Neumann, Josef, 164n22, 178n26
Neumann, Karl Georg, 86
Newton, Isaac, 11, 43, 126, 128; Newtonian physics, 92
nosology, 41, 50, 69, 115, 168n19, 173n63; nosological classification, 99, 168n17
Novalis (Friedrich von Hardenberg), 10, 12, 20, 150, 173n65, 174n72, 182n50; *Das Allgemeine Brouillon* (*The Universal Notebook*), 100; on calculus, 127–8; *Heinrich von Ofterdingen*, 138; on illness, 65–6, 168n18, 176n4; on *Infinitesimalmedicin*, 141, 176n8; on *Lebenskunstlehre*, 109, 176n6; *Die Lehringe zu Sais* (*The Apprentices of Sais*), 56, 67, 151; on oneness with nature, 14, 94, 144–7 passim, 182n1, 183n6; on *Poetik des Übels*, 111; on *Potenzierung*, 140, 154, 180n38, 182n51; on Ritter, 97–8; on self-observation, 82, 96–7; on *Witz*, 54–5

objectivity 7, 25, 32, 94. *See also* subjectivity
Oken, Lorenz, 14, 129, 142, 144, 149; *Lehrbuch der Naturphilosophie* (*Textbook on the Philosophy of Nature*), 125; on *Potenzierung*, 182n51; "Über das Universum" ("On the Universe"), 57
opium, 6, 13, 103, 109, 176nn4–5, 176n9; Brown's dispensing of, 31, 64, 68; Hahnemann on, 13, 28–9, 32–3, 176n4; history of, 162nn6–7
organicism, 49

Packham, Catherine, 117
Paganini, Niccolò, 19, 101
Papsch, Monika, 171n39
Paracelsus (Theophrastus von Hohenheim), 48, 63, 118, 178n23; teaching of signatures, 28, 47, 57, 165n28, 166n41, 169n29
Pfeiffer, Klaus, 85, 176n5
pharmakon, 34, 109, 135
physiognomy, 81, 171n46
Pickstone, John, 45–6, 49, 62
Pilloud, Séverine, 171n37
Pinel, Philippe, 15, 42, 86–7, 172n58
placebos, 41, 112
Platner, Ernst, *Anthropologie für Aerzte und Weltweise* (*Anthropology for Physicians and Sages*), 81
Pollack-Milgate, Howard M., 127
Porter, Roy, 43, 163n10; on classical medicine, 61, 117
Positivism, 93

Potenzierung, 20, 139–41, 154, 156, 160n6, 182n51; Romantics' notion of, 13, 107, 139–40, 157
Pradeu, Thomas, 139
Princess Luise of Prussia, 75–7, 101
Prochaska, Georg, 116
prognosis, 79
prophylaxis, 136, 181n46

Quarin, Joseph von, 14

Reil, Johann Christian, 85–7, 117, 129, 172n56, 176n10; current scholarship on, 172n55; *Rhapsodieen* (*Rhapsodies*), 31, 36, 86; "Von der Lebenskraft" ("On the Vital Life Force"), 44, 120, 123, 169n22, 179n30
Reill, Peter Hanns, 117
Rheinberger, Hans Jorg, 93
Richards, Robert, 117, 177n18, 183n4
Richter, Jean Paul, 21, 146
Risse, Guenther, 12; on Kant, 180n36
Ritter, Johann Wilhelm, 14, 60, 116–17, 123, 130, 142, 144, 149; on analogy, 55–6, 166n37; *Beweis* (*Proof*), 81, 98; as discoverer of ultraviolet light, 11, 126; *Physik als Kunst* (*Physics as Art*), 56, 98; and self-experimentation, 91–7 passim, 125, 139–40; recent studies on, 175n73, 175n75
Rommel, Gabriele, 144, 174n68
Röschlaub, Andreas, 64, 117, 121, 176n9
Rothschuh, Karl, 46, 81, 83, 116, 164n22; on Heinroth, 78, 172n56; on homeopathy, 21, 60, 159n4; on *Krankheitsgeschichten* versus *Krankengeschichten*, 41

Rousseau, Jean-Jacques, 15, 82, 95
Rummel, D., 22, 168n17
Ruston, Sharon, 117

Sachsen-Gotha, Duke Ernst, 15
Sankaran, Rajan, 79
Sarasin, Philipp, 170n32
Sauvages, Boissier de, 37, 42
Scheer, Monique, 83
Schelling, Friedrich Wilhelm Joseph, 8–14 passim, 20, 94, 96, 182n51; and Brown, 22, 64, 168n14; on the living spirit in nature, 125, 129, 142–51 passim, 167n42, 179n32, 183nn6–7; review of Hahnemann, 168n13
Schiller, Friedrich, 20, 85, 94–6, 138, 146; "Über die ästhetische Erziehung des Menschen" ("On the Aesthetic Education of Man"), 137; *Versuch über den Zusammenhang der thierischen Natur des Menschen mit seiner geisten* (*Essay on the Unity between the Animalistic and Spiritual Nature of Man*), 81
Schlegel, Friedrich, 20, 67, 127, 140–5 passim, 175n78, 182n50, 183n6; on the fragment, 100, 175n78; "Lehre von der Analogie" ("Doctrine of Analogy"), 55; on *Potenzierung*, 140–1; on *Witz*, 54–5
Schmid, Carl Christian Erhard, 120
Schmidt, Josef M., 23, 160n7, 161n22, 168n17, 179n28; on Kant, 183n14; "Samuel Hahnemann und das Ähnlichkeitsprinzip," 36, 161n12, 163n8, 163n11
Schopenhauer, Arthur, 120
Schreiber, Kathrin, 65, 98, 113, 170n31

Schubert, Gotthilf Heinrich, 132, 149
Schultz, Hartwig, 139, 143
Schuricht, Ulrich, 78, 112
Schwanitz, Hans Joachim, 168n17
Schwenk, Theodor, 182n2
Sehgal, M.L., 79
Seigel, Jerrold, 61, 67
self-testing, 14, 38, 61, 87, 173n61; Cullen and Störck on, 90, 161n23; Ritter and, 97. *See also under* Hahnemann, Peruvian bark; self-experimentation
semiotics, 36, 41–7 passim, 53–4, 152, 163n13, 165n24
Sertürner, Friedrich Wilhelm, 31
Shakespeare, William, 34–5
Shapiro, Johanna, 6–7
Shelley, Percy Bysshe, 104
Siebold, Georg Christoph, 31
similia similibus curentur, 3, 12–19 passim, 27–57 passim, 65, 68, 99, 145; history of, 163n8. *See also* law of similars
Sloterdijk, Peter, 57, 101–2
Smith, John H., 127, 153
Solhdju, Katrin, 92–3, 174n67
Spinoza, Baruch, 14, 150, 183n5; *hen kai pan*, 142, 147, 152; *natura naturans*, 142, 148, 183n7; *scientia intuitiva*, 142, 152, 183n13
Squirrell, Richard, 134
Stahl, Georg Ernst, 63, 81, 178n21, 178n23; *anima sensitiva*, 118; theory of an *anima*, 42; on therapeutic nihilism, 31
Stalfort, Jutta, 172n51
Stapf, Johann Ernst, 34, 130–3
Steffens, Henrik, 149; on *Potenzierung*, 182n51
Steigerwald, Joan, 116–17, 128, 177n17, 180n37

Stieglitz, Johann, 146, 179n29
Stifter, Adalbert, 138
Stolberg, Michael, 74, 76, 167n1, 167n5, 167n8, 170n31; on fevers, 169n19; on pharmacology, 169n30
Störck, Anton, 14, 44, 90, 164n23
subjectivity, 7, 21, 93, 167n3, 176n9. *See also* objectivity
Sydenham, Thomas, 41
sympathy, 39, 47, 81, 131, 132, 138, 141
Szondi, Peter, 166n34

Tantillo, Astride, 95
Taylor, Charles, 61, 67, 147
therapeutic nihilism, 31, 109, 123
Thoms, Ulrike, 167n6
Thoreau, Henry David, 95
Thums, Barbara, 182n50
Tischner, Rudolf, 127, 174n70, 175n2, 176n9
Tobin, Robert, 174n70
Todorov, Tzvetan, 133
Treuherz, Francis, 22
Treviranus, Gottfried Reinhold, 116–17, 120, 129, 177n18
Tsouyopoulos, Nelly, 65, 168n16
Türk, Johannes, 181n49

Uerlings, Herbert, 65, 176n11

vaccination, 107, 130, 133–8, 146, 181n44
Varady, Helen, 38, 40, 111–12, 161n18
Vila, Anne C., 172n50
Virchow, Rudolf Carl, 15
Vithoulkas, Georgos, 79
Vlasopoulos, Michail, 148
Volkmann, Antoine, 85
Volta, Alessandro, 116, 125

Walach, Harald, 53–4, 58, 106
Waldschmied, Johann Jakob, 31
Walther, Philip Franz von, 117, 121–2
Wedekind, Georg Christian Gottlieb, 146
Wellbery, David, 45
Wettemann, Marion, 39
Wetzels, Walter D., 139–40, 166n37
Wieland, Christoph Martin, 109
Wiesemann, Claudia, 169n20, 176n9
Wiesing, Urban, 64–5, 167n11, 173n63, 175n79, 178n26
Williams, Elizabeth, 119

Wilmans, Carl Arnold, 117, 169n21
Wischner, Matthias, 78, 111–13, 160n6, 169n26, 177n12, 179n31, 179nn33–4, 180n40; and Christian Lucae, 23, 38, 90
Witz, 54–5, 58, 166n35
Wood, David W., 127

Zelle, Carsten, 171n45
Zelter, Friedrich, 95
Ziolkowski, Theodore, 161n21
Zumbusch, Cornelia, 134, 138, 171n47